MATHEMATICS: AIN'T THERE AN APP FOR THAT?

MATHEMATICS: AIN'T THERE AN APP FOR THAT?

AND OTHER FASCINATING MATH ADVENTURES

JERRY FARLOW

THALES HOUSE PRESS

Printed in the United States of America

Drawings by Jean Watts

Published by Thales House Press

First Printing 2016

ISBN: 978-1-5407-6425-6

Contents

Dedication

We've all been there before. We flip to the dedication page in a book and once more, it's been devoted to someone else. Well, this time congratulations are in order. It's to you! Don't worry, there's no need for you to prostrate yourself before me with a thousand thank-you emails, but you might want to tell your friends, relatives, people you meet in the street and others about your unexpected tribute in this insightful, astute, intelligent (and moderately priced) book.

Preface

I was slightly baffled a while back when a student came rushing in my office, but instead of trying to entice me to cough up the blueprints of an upcoming exam or reciting a well-prepared oration for turning in a late homework, she asked a rather curious question.

"Do you remember an old student of yours from 25 years ago?" she said.

Normally, I don't remember the names of students I had the previous semester, but for some reason I remembered this one student.

"Why yes, I remember her," I admitted.

"*She's my mom!*" she blurted out.

And to make matters worse, the incident didn't take place last week. *It was twenty years ago!* So to make sure that another student doesn't come rushing in my office and ask me if I remember so-and-so, then blurt out, "*She's my grandma!*" I decided to take the safe route, retire, and, as they say, get the hell outta Dodge.

—¤¤¤ΞΞ¤¤¤—

A short while after retirement I began to see the benefits of retirement: snoozing in my favorite rocker, lots of relaxation, lots of contemplation, and even more *boredom*. Observing that I was slowly going stark-raving mad, my wife suggested that after teaching and doing mathematical research for 50 years, I might write a book of some kind. Just to verify that I had been doing *something* in the past half-century, she made the wry suggestion.

Little did she know that for the past fifty years, I *had* been writing a book, I just didn't know it. Often, after a long week slaving over a hot differential equations class, I would crash out on Friday night and spew out a chunk of words about some mundane, albeit mathematical, topic.

Although fifty years will result in a pile of words, getting the average book editor, no doubt an English Lit major from a humanities college, to get exercised over such a collection of stories, categorized as "math miscellanea" is a task not taken lightly. Once the average editor sees a book so classified, a rejection letter is not far behind. However, there are a few enlightened editors who see the value of presenting mathematics in various shades of grey, so here we are.

Mathematics: Ain't There an App for That? consists of several short mathematical adventures ranging from satire to serious mathematics to downright silliness.

Feel free to skip over any story that doesn't fit your fancy or contains too much or too little mathematical minutia. Perhaps, however, there are stories which resonate with your curiosity and you will gain something from the experience. Or if you have nothing better to do, feel free to read the entire book, cover to cover. Enjoy.

Jerry Farlow
Professor Emeritus of Mathematics
University of Maine

ΦΣΞΛΘΩ

Forward

The happiest day of my life was the day when this little *tome* was finally shipped off to the printers! Maybe some grass will finally get mowed around this place, or maybe someone, whose name I will not divulge, will take out the garbage!

If I have to answer one more question about whether the period goes before or after the quotation mark, I'll go stark raving mad. Or the difference between "affect" and "effect," sheesh, don't math professors ever learn any grammar?

Now that he's done with his *magnum opus,* as he calls it, he'll no doubt migrate back to his usual headquarters in front of the TV, watching weekend football, and demanding the chips and dips keep coming.

My only desire is that the dear reader of his *tour de force*, gets as much enjoyment from it as I do, knowing it's 100% done.

Susan Farlow
Author's wife

ΦΣΞΛΘΩ

About the Author

My publisher told me that this was the place in the book where I should include all the pretentious crap about myself that I could muster. He said just don't end a sentence with a preposition (or proposition) or mix up commonly misused words like "elicit" and "illicit," but other than that to illicit anything I could think of.

My writing career began on a dark and stormy night when my famed travelwriter wife suggested our funds were trending low, that I might do my part by writing a best-selling [cough] math book.

I said that might be a good idea since my experience in the writing field was established long ago when I spilled a bottle of writing ink on the dress of my first-grade teacher, Miss. Altman, an innocent accident for which she had no quarter. After that came college and my English professor, Mr. Kerrigan, who gave me a D in Eng Comp 101 for my refusal to follow all those fuddy-duddy old rules about composition and English usage.

But things turned around for me after I became a college professor and began accumulating desk drawers of pedagogical tailings. In my attempt to pass on learned words of wisdom to future generations, I spent weekends rummaging through pages of old notes and lesson plans, summarizing their contents in 1,000-word essays. I was ill at ease over the less-than-Harvard-level of scholarship of my writings, and so I published them anonymously in various less-than-Harvard-level publications under the name *Nats Wolraf,* the mirror image of my first and last names, *Stan Farlow*. Although Nat's career as a purveyor of mathematical nuances never reached the stratosphere like those of fellow Mainer,

Stephen King, they did provide motivation for Nats to carry on. Once I got infected with that time-honored habit, I reached out to the textbook field, where no doubt, at this very moment there are legions of students out there poking about in one of my seminal texts on calculus, finite math and partial differential equations, no doubt using my name in vain. Did they actually think the answers to the problems came with the book?

But for those readers who are not into the more serious recesses of the Queen of the Sciences, this book might just tickle their fancy, and do a minuscule amount of educating in the process.

Jerry Farlow

I

MATHEMATICS: AIN'T THERE AN APP FOR THAT?

I drove three days across the blistering Argentine bush before arriving at dusk at a hellhole they call *El Cuidaro.* The town looked like a malignant wart against the Argentine Llanos. As night fell, the streets filled with tough caballeros and desperados. I quickly made my way to a rundown cantina at the edge of town. I entered slowly, an Asus laptop hung from one shoulder, and a myriad of electronic gear across the other. A scratched Freddie Fender record blared mercilessly from a jukebox. The smell of stale beer and cheap booze smacked me across the face.

"Sěnor Farlow," hissed a voice from a table in the corner of the room as a large hairy arm motioned to me. "We've been waiting for you."

I quickly crossed the room and placed my laptop on the table. Sitting down, I flipped back the monitor and hit the power key. A few seconds later, Windows Vista lit up the screen. I quickly moved the cursor to the MS Word 2003 icon when suddenly, I hear snickers coming from around the room.

"You still use Vista?" one man giggled.

"The Gringo still runs Windows Vista," several others cackled like hyenas.

"So what?" I said defiantly. "It does what I want, and I know the commands."

In a few minutes, everyone in the cantina was rolling on the floor, howling with laughter. "He knows the commands," someone said. Having lost everyone's respect, I grabbed my laptop and beat a hasty retreat out the door.

———¤¤¤ΞΞΞ¤¤¤———

Believe it or not, I am trying to make a point here, other than the obvious one that I blew my childhood devouring Louis L'Amour pulp westerns. The story is intended to point out the obvious fact that it's easy to become obsolete in the world of the internet and personal computers. Twenty years ago, college students asked me about computers. Nowadays the situation is reversed.

I hate to say it but my students know more about today's electronics and social media than I do. I'm such a technical luddite that when I want to buy a new tablet, I hang around after class and ask the student in the back row surfing the web for who knows what if I should go for Android or iOS?

Nowadays, if you ask students to prove the Mean Value Theorem, they are likely to ask if there's an App for that, and if you ask them to solve the differential equation , they are likely to write

www.wolframalpha.com/

A student is apt to interpret "do the math," as "find the math" on the internet. The following problems illustrate what students of mathematics in the internet age consider as math solutions.

Problem 1. Evaluate the derivative

$$\frac{d}{dx} x \sin x$$

Answer: Enter der x sin(x) at

www.wolframalpha.com

getting the answer

$$\frac{d}{dx}(x \sin x) = \sin x + x \cos x$$

Problem 2: Graph the function $\sin(t) + \cos(\sqrt{3}\, t)$

Answer: Enter **graph sin *t* + cos (sqrt(3) *t*)** at

www.wolframalpha.com

getting the graph

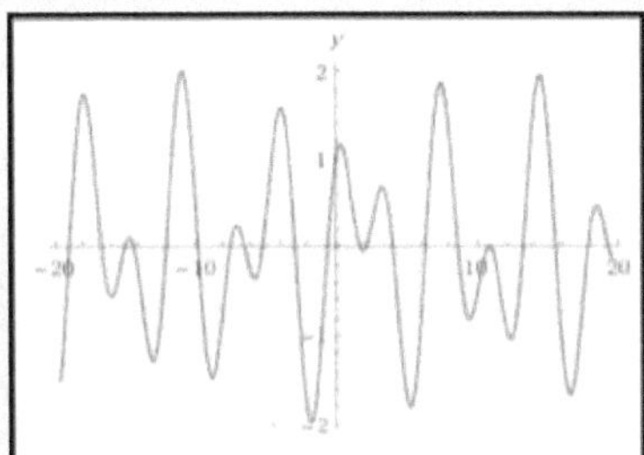

Problem 3: Find the antiderivative

$$\int x \sin x \; dx$$

Answer: Enter **int x sin(x)** at

www.wolframalpha.com

getting the answer

$$\int x \sin x \; dx = \sin x - x \cos x + \text{constant}$$

Problem 4: Solve the initial-value problem

$$\frac{dy}{dx} + y = \sin x, \quad y(0) = 1$$

Answer: Enter solve dy/dx + y = sin(x), y(0) = 1 at

www.wolframalpha.com

getting the answer

$$y = \frac{1}{2}\left(3e^{-x} + \sin x - \cos x\right)$$

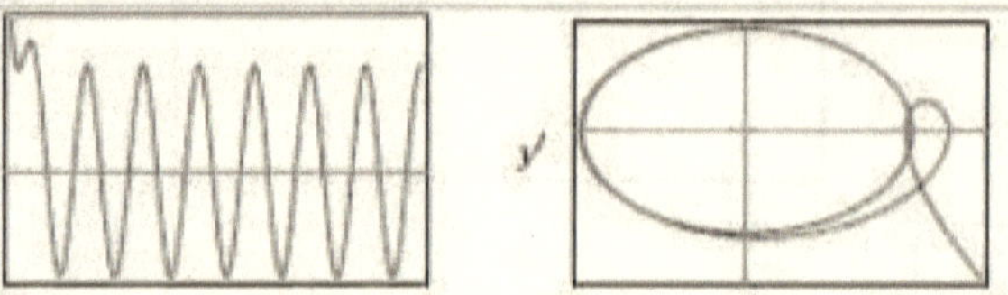

ΠΨΞθΦΧ

2

HOW MATHEMATICS LEARNED TO LOVE THE X

When someone asks me what do I do for a living, I tell them I'm trying to find x. Sometimes I even succeed, as when someone asks me to find the x in the equation

$$x - 1 = 0$$

or in my youth in beginning geometry class when I exhibited a mastery of the Pythagorean Theorem.

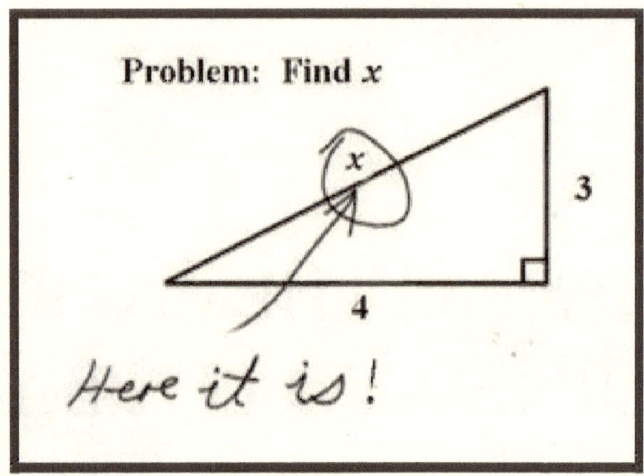

Hmmmmmmmmmm

But mathematicians are not the only people fascinated with the letter x. The 24th letter of the alphabet has been used for everything from "marking the spot" to warning us about "dirty movies." But the real mystery is, why do mathematicians use x? Why aren't we ever asked to find p in the equation

$$p - 1 = 0$$

or warn us about p-rated movies?

One theory is because the Spanish can't say "sh." A thousand years ago the root of mathematical learning was centered in the Islamic world, where the first systematic study of algebra was starting to percolate. When Arab mathematicians solved an equation for a certain value, they referred to the certain value as "something," which in Arabic is written as

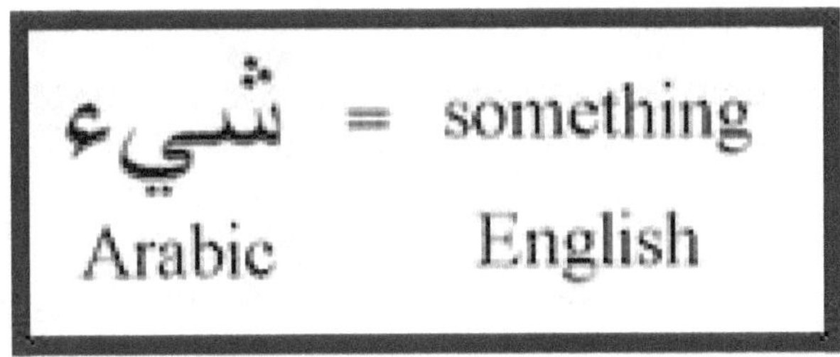

Arabic pronounced "sh"

and pronounced as "sh." When Arab mathematical texts arrived in Spain at the beginning of the European renaissance, scholars translated these works into Spanish. The problem they faced was that the Spanish language does contain a sound for "sh" so the translators went with the "ck" sound, which in Greek is closest to the letter chi, written as χ. But European writers of math books preferred Latin letters to Greek letters and so the Greek χ morphed into the Latin *x*.

And like they say, the rest is history.

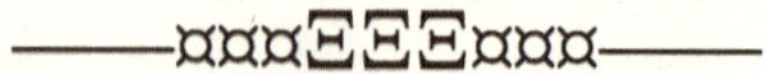

The theory of how math discovered *x* has some similarities to how Christmas is sometimes expressed as Xmas — religious scholars used the Greek letter chi as shorthand for Christ.

Nowadays, it is fashionable to steal the mathematical *x*, which is understood to stand for the "unknown" or "mystery." There are X-rays, which were invented in the 1890s, and at the time were of unknown origin. Then there were the black Muslims, taking a new unknown name as a replacement for slave names, the most famous being Malcolm X. And don't forget about all the Generation X cohorts, who are essentially Baby Boomers who lie about their age.

The *Los Alamos National Laboratory* even has a super-secret division, called the X Division, where some of our nation's top eggheads are responsible for the safety and reliability of the nation's nuclear deterrent. One wonders about the intelligence of using X to designate its most classified area inasmuch as any spy with any intelligence who managed to sneak into the place would head right for the X-files and not waste time on the A-files.

———¤¤¤ΞΞΞ¤¤¤———

But there are other theories about how mathematics fell in love with the *x*. The problem with the above theory of

Arab → Spanish → English

is firstly, there is no documented evidence to support it, and secondly, medieval scholars who translated these works cared little about the phonetics of words but on their meaning.

For another and more documented version of the story of *x*, we turn to the German philosopher and mathematician Renè Decartes, who used *x* in his 1637 seminal work *La Géométrie.*

It is uncertain whether Descartes originated the idea for using *x* as a practice for representing an unknown quantity or borrowed it from someone else. Nevertheless, as far as documented evidence goes, he seems to be the creator, if not the person who popularized its use, as noted by the math historian Florian Cajori in his book *A Theory of Mathematical Notations.*

In *La Géométrie,* Decartes solidified the mathematical notation we use today with *a*, *b* and *c* designating *known* quantities and *x*, *y* and *z* designating *unknown* quantities.

ΠψΔΦθΛ

3

MATH ANECDOTES AND URBAN LEGENDS

The history of mathematics is replete with fascinating anecdotes about famous mathematicians from the past — and as with most remembrances, they range from being completely 100% accurate to 100% phony urban legends at the other. More often than not, however, they belong to the true-but-embellished category.

The story of how the sickly French mathematician Rene Decartes, who spent much of life in bed thinking about mathematics and philosophy, noticed a fly crawling around on the ceiling. Realizing he could tell others the location of the fly by its distance from the walls of the room, he got out of bed and wrote down his ideas and in the process invented the coordinate plane, which we now call the Cartesian plane. Whether this story is true or simply legend no one knows, but it has been told over and over for four hundred years.

On the other hand, the story of how George Dantzig solved previously unsolved problems, mistaking them for homework, is verifiably true. And don't forget the anecdote about how the Swiss mathematician Leonard Euler humiliated the great French atheist philosopher Diderot when Diderot was unable to answer Euler's absurd mathematical proof of the existence of God:

"Monsieur

$$(a + b^n)/n = x$$

therefore God exists."

It may be an intriguing story although it's been thoroughly debunked by numerous historians.

Although anecdotes about a discipline give the subject a sense of place, a connection with the culture, it takes just one lone biographer willing to stretch the truth in pursuit of literary entertainment to destroy any resemblance of historical correctness. The writers that follow, often too lazy to check the facts [cough] and adding embellishments [cough cough] of their own, add to the confusion. The process of arriving at the final anecdote can be compared to the childhood game of "telephone," where one child whispers a message into another child's ear, who in turn whispers it to another child and so on until the last child hears the message and recites it out loud for all to hear: "*The cat has three tails except on Friday when John never went to school.*"

The following anecdotes from mathematics past belong either to the 100% accurate category, to the true-but-embellished category, to ... well, in the case of Einstein's driver, one of the best phony urban legends out there.

Dantzig's Homework Assignment

In 1939, George Dantzig was a brilliant doctoral student at the University of California, Berkeley. Upon arriving late for a statistics class, he discovered two problems written on the blackboard. His professor, the eminent statistician Jerzy Neyman, had written down these problems for the class, being examples of unsolved problems in statistics. Dantzig wrote them down, thinking they were assigned problems — and as any good student would do, he solved them.

Upon turning in the solutions, Dantzig apologized for turning them in late since they seemed a little more difficult than normal. A few weeks later Neyman woke Dantzig and his wife on a Sunday morning at eight o'clock, telling Dantzig that he, Neyman, had just written an introduction to a paper detailing one of Dantzig's "homework solutions", on which Neyman had written Dantzig as the author.

Later when it came time for Dantzig to write his Ph.D thesis, Neyman told him to simply put the two solutions in a binder and he would accept them for the thesis.

PS: This anecdote is 100% true as retold by Dantzig himself. The two papers Dantzig published, thinking they were homework problems were the following:

- *On the Non-Existence of Tests of 'Student's' Hypothesis Having Power Functions Independent of Sigma* by George Dantzig. Annals of Mathematical Statistics. No. 11; 1940 (pp. 186-192).
- *On the Fundamental Lemma of Neyman and Pearson* by George Dantzig and Abraham

Wald. Annals of Mathematical Statistics. No. 22; 1951 (pp. 87-93).

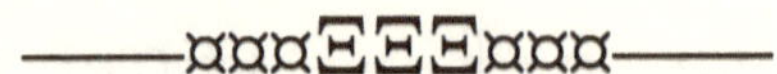

von Neumann and the Flying Fly Problem

There is a problem with which every student of calculus is familiar. Two trains, 200 miles apart are heading towards each other on the same track, each traveling at 50 miles per hour. A fly starts at one train and flies 75 miles per hour towards the other train, whereupon it turns around and flies back to the first train — whereupon it turns around again and goes back to the other train. It continues this process of flying back and forth between the trains until SQUASH when the trains collide. The problem of this morbid tale is to determine the total distance that the fly flies.

Since the fly flies back and forth between the trains an infinite number of times, each distance getting smaller and smaller, the calculus student can solve the problem by summing the infinite series for all the back-and-forth distances. However, a far easier solution is to observe that since the trains are 200 miles apart and each travels 50 miles per hour, it takes two hours for the trains to collide. Hence, the fly traveling at 75 miles per hour flies a grand total of 150 miles. It is as simple as that.

When the problem was posed to the great mathematician John von Neumann, known for his prodigious computational ability, he replied immediately, "150 miles."

"It is a very strange problem," observed the poser. "Most people try to solve the problem by summing the infinite series."

"What do you mean strange?" replied von Neumann. "*That's how I did it.*

Lecturer X's Lecture on Fermat's Last Theorem

A famous mathematician once announced that he would be presenting a proof of Fermat's Last Theorem at a conference, but when he gave the lecture, he didn't once even mention the famous theorem, let alone prove it. After the lecture someone in the audience asked him why he said he'd prove Fermat's Last Theorem but didn't even mention it. The lecturer replied he did it just in case he was killed on the way to the conference.

Grade School Gauss Nails

$$1 + 2 + \cdots + 99 + 100 = 5050$$

Carl Friedrich Gauss (1777-1855) is considered one of the greatest mathematicians of all time, making important contributions to almost every area of mathematics, as well as physics, astronomy and statistics. And like many of the greats, he showed amazing talent at an early age.

One childhood story tells the tale of when Gauss was in primary school. One day Gauss' teacher asked the class to sum the numbers from 1 to 100, assuming the task would occupy them for quite a while. The teacher was shocked when young Gauss, after only a few moments, told the teacher the correct answer of 5050. The teacher was puzzled how an eight-year old child could find the answer so quickly. Young Gauss told the teacher the problem was really quite simple. He simply added numbers from 1 to 100 in pairs – the first and the last, the second and second to last and so on, observing that

... then, adding these 50 values of 101 gives

first pair: $1 + 100 = 101$
second pair: $2 + 99 = 101$
third pair: $3 + 98 = 101$
... ...
fiftieth pair: $50 + 51 = 101$

arriving at 50 values of 101. Hence, the answer is $50 \times 101 = 5050$.

Johann Bernoulli and the Brachistochrone Problem

The invention of calculus in the later 1600s, as great a discovery as it was in the history of mathematics, brought forth a contentious battle between the followers of the two co-inventors of the subject over who should receive the accolades for its discovery. Both co-inventors, the English mathematician/physicist, Isaac Newton (1642-1728), and German mathematician/philosopher Gottfried Leibniz (1646-1716), each had their devoted disciples.

Two of the more colorful enthusiasts of Leibniz were the Swiss mathematician Jakob Bernoulli (1655-1705) and his younger brother Johann (1667-1748), who regarded Leibniz as the true founder. Although both brothers were extremely competitive and cantankerous, they were brilliant mathematicians and played a major role in the early development of calculus.

The story of how Johann Bernoulli will forever be linked to Newton is legendary in mathematical lore. The story goes that in 1696 Johann Bernoulli published a problem in Leibniz's mathematical journal *Acta Eruditorum*, where he challenged the mathematical community to solve the following problem, which he called the *Brachistochrone Problem*, stated as

... two points A and B lie at different heights above the ground with one point not lying directly above the other. There are an infinite number of curves connecting the two points, ranging from a straight line to an arc of a circle. A bead sliding down a curve takes a certain length of time and the time depends on the shape of the curve. The problem is to find the shape of the curve that minimizes this time.

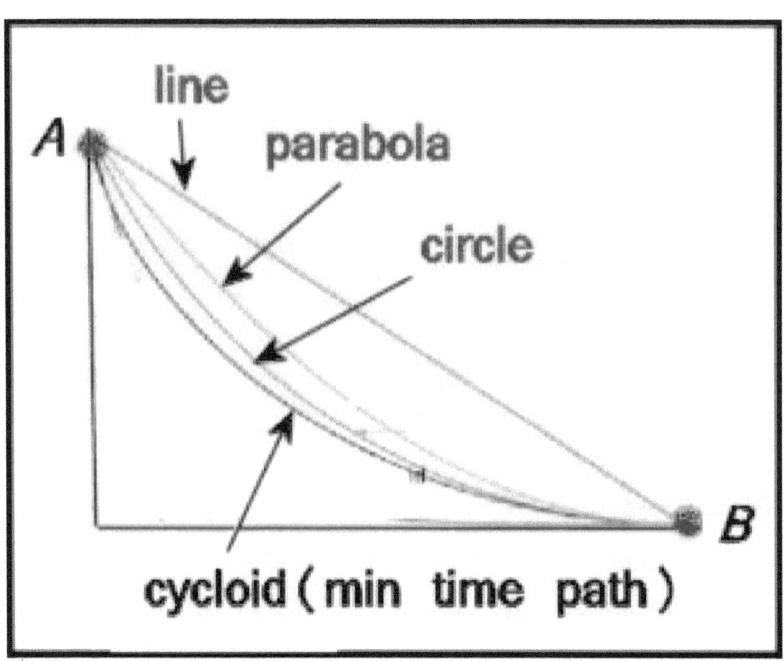

Brachistochrone problem

Johann Bernoulli himself had solved the problem, a fact he did not hide from the mathematical community. He set a deadline of six months for solutions to be submitted, and when the deadline came one person had returned a correct solution — that being the great Gottfried Leibmiz. At the suggestion of Leibniz, Bernoulli decided to extend the deadline for solving the problem and in the process made a veiled reference to the fact that "certain persons" had not managed a solution — no doubt referring to Newton. He wrote

> *... so few have appeared to solve our extraordinary problem, even those who boast that through special methods they have not only penetrated the deepest secrets of geometry but also extended its boundaries in marvelous fashion; although their golden theorems, which they imagine known to no one, have been published by others long before.*

It was clear the message was intended to deride Newton, whose method of "fluxions" was considered by Newton's followers the gold standard for solving such problems. By this time, Newton was no longer actively involved with the calculus, having co-invented the subject in the 1660s, thirty years earlier. Bernoulli, wanting to leave no doubt that Newton was aware of the problem, sent him the problem by mail.

The story is then told by Newton's niece, who relates the story that Newton had just returned home, exhausted from a long day at the Mint where he was involved with the great re-coinage. Upon reading Bernoulli's letter, Newton felt his honor and reputation were being challenged and sat down and solved the problem within a few hours. He is said to have remarked, "*I do not care to be teased by foreigners about mathematical things.*" Back in Europe when the extended deadline arrived, Johann Bernoulli had five correct solutions, the correct curve being an *upside cycloid*, where a cycloid is the curve traced by a point on a rolling wheel.

In addition to his own solution, the other persons who correctly solved the problem were his brother Jacob Bernoulli, Gottfried Leibniz, the

Marquis de l'Hôpital, and an anonymous, unsigned letter with an English postmark.

Legend has it that Bernoulli, although an outward detractor of Newton, nevertheless harbored huge admiration for the great genius and the story goes he said, "*I recognize the lion by his paw.*"

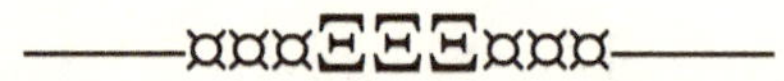

Einstein's Chauffeur

When the greatest physicist of the twentieth century, Albert Einstein immigrated to the United States in 1933 to join the Institute of Advanced Study at Princeton University, he often lectured at various universities to enthusiastic audiences. Although few people knew much about the famous scientist and even less of what he was talking about, Einstein felt it was his obligation to create an interest in science. Since Einstein never learned to drive, he was given a chauffeur, Harry, to accompany him on his travels. Although Harry had no knowledge of mathematics and science and didn't understand a single word of Einstein's lectures, he attended every one, sitting attentively in the back row.

One day on the way to a lecture, Einstein became ill and told Harry he would be unable to give the day's lecture. Harry addressed him and said, "Professor Einstein, I've heard your lecture a hundred times and although I do not understand it, I can nevertheless deliver it word-for-word, the exact same way you have always done." Einstein thought the idea absurd but after realizing that no one at the current university knew him by sight, he reluctantly agreed on Harry's unusual suggestion. Einstein told Harry that he would sit in the back row during the lecture.

Before the lecture, a cocky professor was boasting to fellow colleagues about his expertise on Einstein's theory of relativity. He even bragged that there were aspects of the theory he probably had a deeper understanding of than the great physicist himself. "After the lecture, I'm going to ask Einstein some questions to see if he can answer them," he continued.

The lecture then began and as promised Harry delivered it brilliantly, not a word or phrase out of place — all the while Einstein sitting in the back row, wearing a chauffeur's uniform and playing his new role as "Einstein's driver".

After Harry finished the lecture but before he could leave the speaker's platform, the boastful professor raised his hand and asked his prepared question about Einstein's theories, one which involved complicated equations and formulas, intending to put his expertise of the subject on display. Harry flinched for a moment but then regained his composure and said

> *"A fine scholar you are. I cannot imagine anyone could ask such a simple question. As a matter of fact to show you how simple it is, I'm going to let my driver at the back of the room answer it."*

ΠψΔΦθΛ

4

FOR STUDENTS WHO SLEPT THROUGH ALGEBRA 101

No doubt Alice and Bob were half asleep when their Algebra 101 teacher informed the class that when you raise a number to a power, like 2^3 and then raise that number to another power, like

$$(2^3)^2$$

the order in which you raise the powers makes no difference, as in

$$(2^3)^2 = (2^2)^3 = 2^6 = 64$$

Although this fundamental algebraic property may not raise many a brow amongst beginning algebra students, it does play a central role in cryptography, in particular the Diffie-Hellman Public Key Exchange.

In 1976 Whitfield Diffie and Martin Hellman introduced a clever but powerful way for two computer users to create a shared private key (i.e. large number) which they can use as their secret "password" to code and decode private messages across an insecure network.

———¤¤¤ΞΞΞ¤¤¤———

The Diffie-Hellman algorithm depends on the aforementioned exponential rule as well as modular (or clock) arithmetic. A good way to think about modular arithmetic is to think about integers 0,1,2,... arranged around a circle an infinite number of times instead of on the number line.

Figure 1 shows a clock (i.e. circle) consisting of seven hours labeled 0,1,2,3,4,5, and 6. Going around the clock a second time the hours are labeled 7,8,9,10,11, 12, and 13. Going around again they're 14,15,16,17, 18. 19, and 20, and so on forever. In modular arithmetic (mod 7) we take the numbers 0,7,14,21,...to be the same number but we don't say "equal," we say they are *congruent* mod 7.

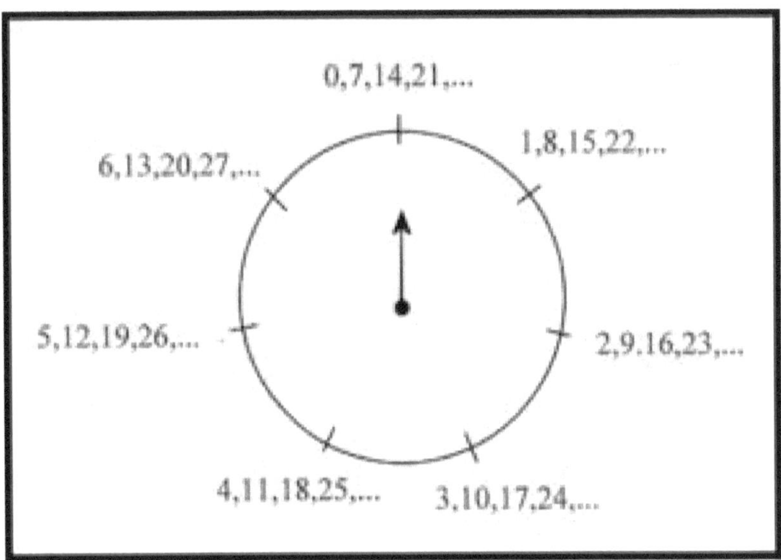

Figure 1: Telling time mod 7

Also note that in the 7-hour clock all the congruent numbers at each "hour", like 1,8,15,22,... have the same remainder, in this case 1, when divided by the modulus 7. You can check that the other "hours" of the clock also have the same remainder when divided by 7. Keeping time on a mod 7 clock is similar to the way we tell time on our mod 12 clock, which we use all the time.

In order to add 3 plus 6 in the above mod 7 clock, we start at 0, count (always clockwise) to 3, then another 6 arriving at 2. In other words $3 + 6 = 2$ in mod 7 arithmetic is 2, which we write as

$$3 + 6 = 2 \bmod 7$$

To multiply 4 times 3, start at 0 and rotate 4 hours around the clock 3 times, arriving at 2, thus $4 \times 3 = 5$ in mod 7 arithmetic, or simply

$$4 \times 3 = 5 \bmod 7$$

Now for the main event: the Diffie-Hellman Key Exchange algorithm.

Diffie-Hellman Key Exchange

The Diffie-Hellman Key Exchange algorithm

addresses the problem of two parties sharing a secret number (Exchange Key) used to encode and decode private messages when communications are carried out over a public channel that can possibly be intercepted by unwanted parties.

Alice and Bob wish to transmit personal messages over the internet and need a secret key (i.e. large number) to encode and decode their messages so the notorious eavesdropper Eva is unable to read them.

Here is what they do — wake up Algebra 101 students.

Public Modulus and Generator: Alice and Bob agree on two prime numbers p and g that have no common factor. The number p is called the *modulus* and g the *generator*. It is not necessary to hide these numbers from Eva or any other evesdropper for that matter. In order for the process to be effective however, the modulus p must be very large, like a *300 digit number.* The generator

g does not have to be large and can even be the smallest prime number 2.

In our example, however, Alice and Bob agree on a tiny modulus of $p = 7$ and a generator of $g =$ 3.

———¤¤¤ΞΞΞ¤¤¤———

Alice's Secret Number: Alice now picks a number a she calls her *secret number*, known only to herself, and she decides on $a = 4$.

Bob's Secret Number: Bob picks a number b he calls his *secret number*, known only to himself, and he decides on $b = 2$.

———¤¤¤ΞΞΞ¤¤¤———

Alice's Public Key: Alice now computes what she calls her *public key* by the modular arithmetic formula

$$A = g^a \bmod p = 3^4 \bmod 7 = 81 \bmod 7 = 4$$

Bob's Public Key: Bob now computes his *public key* by

$$B = g^b \bmod p = 3^2 \bmod 7 = 9 \bmod 7 = 2$$

Public Key Exchange: Alice and Bob now exchange their public keys over the insecure network. They are unconcerned that Eva might be reading Alice's public key of 4 and Bob's public keys of 2.

Computing the Private Key: Alice and Bob are now in position to compute their common *private key*, a key that Eva will not know..

Alice now takes Bob's public key of $B = 2$ and using her private number $a = 4$ and computes

$$P = g^{aB} \bmod p = 3^8 \bmod 7 = 6561 \bmod 7 = 2$$

Bob now takes Alice's public key of A = 4 he received from Alice and using his secret number of $b = 2$, computes

$$P = g^{Ab} \bmod p = 3^8 \bmod 7 = 6561 \bmod 7 = 2$$

which is exactly the same private key computed by Alice. Alice and Bob can now use this private key of 2 to encode and decode messages. Note that Eva has no idea of the value of this number. In this example the private key is small, but in real situations when the modulus *p* is a 300 digit number, the private key will be is upwards of 300 digits.

And to think the method depends in great part on the simple exponential rule

$$(g^a)^b = (g^b)^a = g^{ab}$$

ΠΨΞθΦΧ

5

THE PROBLEM WITH MY DEAR AUNT SALLY

There's the story about the girl in a beginning writing class who tells the teacher that the phrase

a woman, without her man is nothing

lacks a comma, whereupon she writes

a woman, without her, man is nothing.

Other than the fact that the teacher is probably male and the student female, this story brings up the fact that although grammar is important for precision in natural language, precise rules are also important for making arithmetic unambiguous. For example, what number does the expression

$$1 + 2 \times 3$$

represent? Some people say 9, others say 7. In other words, which of the two equations is the correct interpretation

- $(1 + 2) \times 3 = 3 \times 3 = 9$
- $1 + (2 \times 3) = 1 + 6 = 7$?

It is a commonly accepted rule that when performing arithmetic operations involving addition and multiplication, one performs multiplication before addition, hence, the mystery expression normally represents the number 7. The convention that multiplication precedes addition was used in the earliest textbooks dating back to the 16th century.

If you had been paying attention in grade school, you might recall your teacher telling you the easily-remembered PEMDAS rule:

Please Excuse My Dear Aunt Sally

which roughly governs the order in which arithmetic operations should be performed. That is

1. first do operations inside Parentheses
2. then evaluate the Exponents
3. then do the Multiplications
4. then do the Divisions
5. then do the Addtions
6. finally do the Subtractions

Using these rules Aunt Sally would interpret

- $1 + 2 \times 3 = 1 + 6 = 7$
- $3 + 4/2 = 3 + 2 = 5$
- $(3 \times 2) + 2^3 + 1 = 6 + 2^3 + 1 = 6 + 8 + 1 = 15$

However, what your grade school teacher failed to say is that there are no universally accepted rules for the order in which arithmetic operations should be performed. Some of the early four-key calculators without a stack implement chain input did not obey the PEMDAS rules, but took the easy route and carried out arithmetic computations from left to right without regard to any specific rules. Such a calculator would compute

$$1 + 2 \times 3 = (1 + 2) \times 3 = 3 \times 3 = 9$$

Adhering to the PEMDAS rules, which state that multiplication precedes division, one would compute

$$6/2 \times 3 = 6/(2 \times 3) = 6/6 = 1$$

whereas the majority of people (you too?) assume the 3 is in the numerator following the quotient 6/2 and hence would compute

$$6/2 \times 3 = (6/2) \times 3 = 3 \times 3 = 9$$

In fact, the most commonly accepted rules for ordering arithmetic operations that contain division/multiplication and addition/subtraction is that there is no order of preference but to carry out computations from left to right as described by the following rules.

1. First do all operations inside parentheses.
2. Next evaluate all exponents.
3. Next do multiplications and divisions from left to right.
4. Finally do additions and subtractions from left to right.

There is nothing sacrosanct about these rules. They are simply a convention for consistency in arithmetic computations. The origin of the rules is unknown, whether one person was responsible, or whether they were adopted gradually by the mathematics community, is unknown. That said, many early textbooks used different rules and hence got different answers for the same expression. The fact that mathematical notation for the arithmetic operations +, /, ×, and – has changed over the centuries no doubt has added to the confusion. However, today's textbooks tend to agree that multiplication/division as well as addition/subtraction are not carried out in any particular order but should be performed from left to right.

An interesting exercise for the reader would be to enter various arithmetic expressions at online websites that carry out arithmetic operations. You will discover that some websites yield different answers for the same arithmetic expression.

Algebraic operations are sometimes computed differently depending on the calculator. The algebraic expression 1/2x was interpreted as 1/(2x) by the Texas Instrument TI-82, but as (1/2)x by the TI-93 and every TI calculator since 1996. The moral is, when you are in doubt about the meaning of an expression, add parentheses!

Computer languages also vary on the order in which they perform arithmetic operations. Most languages (Fortran, C, C++, ...) carry out computations in the order 1,2,3,4 described above, although some languages like *APL*, *Smalltalk* and *Occan* carry out their computations from left to right. In fact, the computer language *LISP* has *no rules at all*, the programmer must specify the order in which the computations should be performed.

Computer languages handle arithmetic statements by converting them to data structures called trees by programs called parsers. For example, the parsed tree for the arithmetic statement $(5 + 2) \times 7$ is

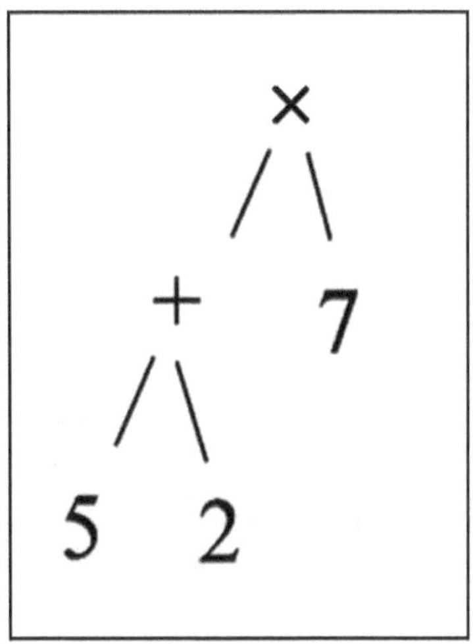

Parsed tree for (5 + 2) x 7

Parsed trees are useful since all arithmetic (and computer language as far as that goes) is converted to a standardized format from which they can be manipulated and evaluated by formalized rules. This is the computer way of doing arithmetic.

ΘΠΩΛΦΨ

6

CAN YOU SOLVE THESE? BRING IT ON

Even at the college level, I lay down strict rules for personal conduct on the last day of class: no throwing books in the air, no running in the hallways, and no fist-bumping or high-fives to everyone you meet — at least not until the students are gone.

Although I have always honored the aspirations

of my students, they have, on occasion, caused me to browse the faculty ethics manual, researching permissible grounds for student strangling. One student activity which has always chafed a few tail feathers is the practice students missing class due to a weekend of carousing and then materializing in my office and asking me nonchalantly if he or she missed anything at all important. I say of course not, we didn't introduce the Fundamental Theorem of Algebra, we spent the hour watching cat videos on YouTube.

On the other hand, more than one student has told me there are professor mannerisms which cause animus among the student body. Introducing new material a minute before the bell rings, or asking trick questions on an exam come to mind.

Now I can understand objections to introducing new subject matter a minute before closing time, but trick questions? Why, those little gems of tutelage are my hallmark! Just ask any of the spoilsports who post those anonymous threats on my message board.

To present my case on the fairness of my bet-you-can't-get-this-one problems, I have dug out a few of my favorite stumpers to test your mettle. Good luck.

Problem #1 [Harry's Biased Penny]

People have been flipping coins (or what we in mathematics call Bernoulli trials) for centuries for the purpose of everything from resolving arguments, to who receives the football on kick-offs, to giving the city of Portland, Oregon its name in lieu of Boston, Oregon. However, coins are not as unbiased as we are lead to believe. Some researchers at Stanford University have determined there is a 51% chance the side facing upon landing is the same side facing up when flipped. The question then arises, how can you mastermind a fair coin from an unfair one — and we pose the question to you.

For example, suppose John has a biased coin which turns up heads 75% of the time and tails 25% of the time. We denote the probability that the coin turns up heads by p and tails by q. How can John take this unfair coin to create a process that turns out one way 50% of the time (which we can call heads) and the opposite way 50% of the time (which we call tails)? Hmmmm?

We let you ponder your own strategy.

To make a fair coin, John flips his biased coin two times. The result will be one of four possible outcomes HH, HT, TH, and TT as displayed in the following drawing, where H represents turning up heads and T represents turning up tails.

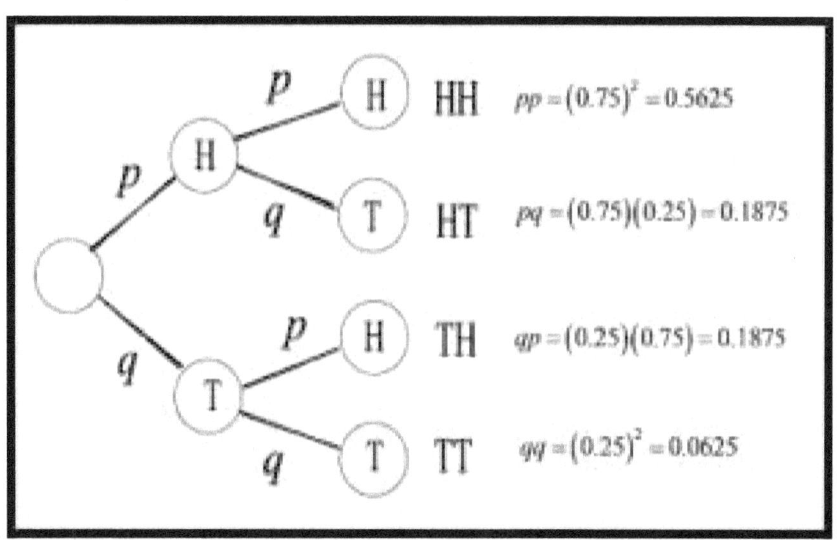

Graphical illustration of flipping a coin two times

The probably of getting two heads (HH) is the product of getting a head on the first toss times the probability of a head on the second toss, or

$$pp = (0.75)^2 = 0.5625$$

The probability of getting the other options HT, TH, and TT are computed in a similar way; that is, the probability of getting a head or tail on the first toss times the probability of getting a head or tail on the second toss.

Note that the probability $pq = 0.1875$ of getting a head on the first toss and a tail on the second toss is the same probability $qp = 0.1875$ of getting a tail on the first toss and a head on the second toss. Hence, we can devise a strategy as illustrated in the following drawing.

The strategy for "constructing" a fair coin is now clear. You simply flip your unfair coin two times. If both tosses turn up heads or both tails, forget it and toss two more times until the outcome is either a head followed by a tail or a tail followed by a head. If the outcome is a head followed by a tail, call this outcome a HEAD, and if the outcome is a tail followed by a head, call this outcome a TAIL. The probability of getting a HEAD is the same as getting a TAIL — a fair coin. Neat, eh?

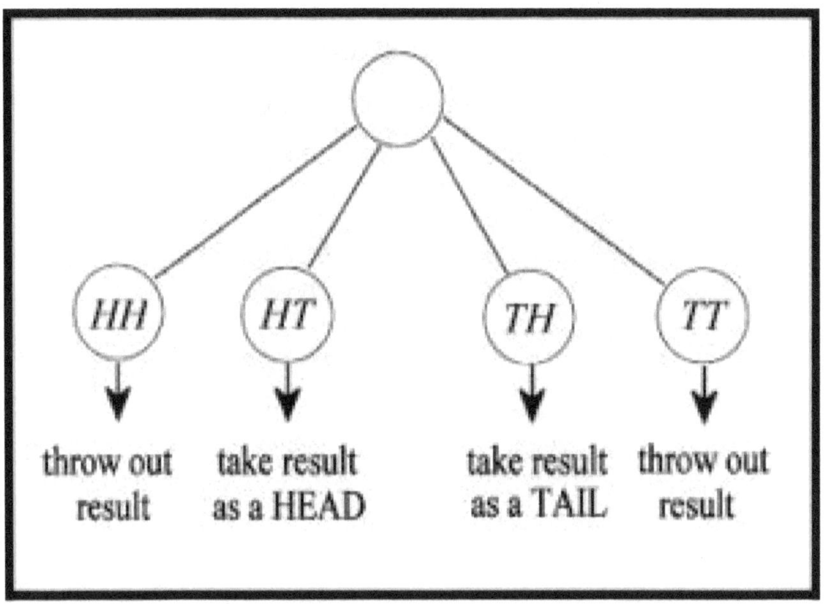

The solution revealed

You still want another one? This one is a little more serious, but thought provoking.

Problem #2 [Mary's Medical Screening]

Consider a medical screening study to test a certain population for a disease. The test for the disease is at least 95% accurate, meaning that if the person has or does not have the disease, the test gives the correct result at least 95% of the time. Suppose Mary is taking part in the study but unfortunately her test comes back positive. Does this mean there is at least a 95% chance that Mary has the disease?

You can answer the question now.

You say yes? The answer is no, not necessarily. The reason is because there's a big difference between the two *opposite conditional* probabilities:

1. probability the test is positive given the person is sick
2. probability the person is sick given the test is positive

Suppose the results of the study which Mary participates is shown in the table below. The rows determine the accuracy of the test while the columns are what Mary and others participants are more concerned.

	Test +	Test –	Total
Sick	5 true +	0 false –	5
Healthy	25 false +	970 true –	995
Total	30 test +	970 test –	1000

Row-wise: The first row of the above table shows that of the five individuals who have the disease, the test detected the disease all five times for a 100% accuracy. For the 995 non-diseased (healthy) individuals displayed in the second row, the test was accurate 970 times for a 95% accuracy. Hence, regardless of whether a person has or does not have the disease, the test is accurate at least 95% of the time.

Column-wise: However, Mary is not concerned with looking at the table row-wise, but column-wise, in particular the first column. The first column says that of the 30 individuals, like herself, who tested positive, only 5 persons actually have the disease, which means the probability that she actually has the disease is 5/30 = 0.16 (16.6%) and not the 95% as one might think.
The reason for this relatively small probability is due to the fact that most of the positive test

results came from non-diseased individuals since most people in the study fall in the non-diseased category.

ΠΨΞθΦΧ

7

IS IT MUSIC OR IS IT MATHEMATICS?

In the category of "please hit me over the head with a hammer," one cannot help but be amazed about a Jason Padgett, who in 2002 was attacked and beaten by two men outside a karaoke bar. He suffered a severe concussion but recovered and surprisingly the incident rattled some brain cells, unlocking a previously dormant portion of his brain and turning him into something the rest of us can only dream, a mathematical savant, able to see the world through the lens of geometry. He belongs to a rare collection of individuals who have developed prodigious abilities after a severe injury or disease.

"I see shapes and angles everywhere," he told his doctor.

I, myself, have been rapped on the cranium more than once, at least vicariously by a student now and then. They are not of the severity to transform me into a topological genius, but sturdy enough so that I've developed *Mathematical Synesthesia,* a rare malady that causes me to see mathematics where others see only fields of blossoming daisies.

My own theory is that most people suffer from *Mathematical Synesthesia* in varying degrees. Some of the following quotes are about *music, art*, and *poetry,* while others are about **mathematics**. It is the reader's task to determine which quotes refer to **mathematics** or to the more artsy topic indicated by the word in *italics.* The author of each quote is included which provides a clue, but keep in mind that many famous artists, poets, musicians, and other thinkers have strong opinions, positive and negative, when it comes to mathematics, so think before you decide. An interesting observation of this experiment is the wide variety of responses among readers. The answers to this little "math is everywhere" interrogation are given at the end of the quotes. Don't peek.

———¤¤¤☰☰☰¤¤¤———

1. ["*Music*, **Mathematics**] expresses that which cannot be put into words and that which cannot remain silent." —Victor Hugo
2. ["*Art*, **Mathematics**] is the lie that enables us to realize the truth."
—Pablo Picasso
3. ["*Poetry* **Mathematics**] strikes the reader as words of his own highest thoughts." —John Keats
4. "My [*religion*, **mathematics**] consists of a humble admiration of the illimitable superior spirit who reveals himself in the

slight details we are able to perceive with our feeble mind." —Albert Einstein

5. ["*Biology*, **Mathematics**] is the study of complicated things that have having been designed with a purpose." — Richard Dawkins
6. ["*Love*, **Mathematics**] looks not with the eyes but with the mind." —William Shakespeare
7. The best part of [*beauty*, **mathematics**] is what no picture can express." — Francis Bacon
8. ["*God*, **Mathematics**] has no religion." — Mahatma Gandhi
9. "If you *read something you can't understand, you can be sure that it was drawn up by a [lawyer,* **mathematician**]." — Will Rogers
10. "I tell them that if they will occupy themselves with the study of [*literature*, **mathematics**] they will find in it the best remedy against the lusts of the flesh." — Thomas Mann
11. ["*Golf*, **Mathematics**] is simple but endlessly complicated." — Arnold Palmer
12. "For the things of this world cannot be

made known without a knowledge of [*philosophy*, **mathematics**]." – Roger Bacon

13. "If [*God*, **mathematics**] did not exist, it would be necessary to invent it." – Voltaire
14. "Whoever wishes to become a [*philosopher*, **mathematician**] must not be frightened by absurdities." –Bertrand Russell
15. ["*Sex*, **Mathematics**] lies at the root of life and we can never learn to revere life until we understand it." —Havelock Ellis
16. "We may not pay [*Satan*, **mathematicians**] reverence, although we respect its talents." —Mark Twain
17. "What is [*hell*, **mathematics**] ? I maintain it is the suffering of being unable to love." —Fyodor Dostoyevsky
18. ["*Life*, **Mathematics**] is like riding a bicycle, keep moving." —Albert Einstein
19. ["*Philosophy*, **Mathematics**] is the most sublime of human pursuits." — William James
20. "Chicks dig [*historians*, **mathematicians**]." — Cynthia Hand

ΠΨΞθΦΧ

Answers: The quotations numbered 5, 10, and 12 are about mathematics, the others are about the alternate word in *italics.*

8

THE MOST INTERESTING MATHEMATICIAN IN THE WORLD

He once made a math error on purpose just to experience how it felt. He won a lifetime achievement award for his groundbreaking contributions to mathematics — three times. People who suffer from math anxiety sit at his feet to learn mathematics. If he fails to solve a mathematics problem, it's insolvable. He doesn't always do math, but when he does he receives a Field's Medal. He is ...

> *the Most Interesting Mathematician in the World.*

So, apart from the suave, bearded Jonathan Goldsmith, the iconic actor in the tongue-in-cheek commercials for *Dos Equis* beer, who exactly is ... *The Most Interesting Mathematician in the World?*

The jury is still out since mathematicians come in all shapes and sizes. Many of the great ones had eccentric and prickly personalities, others are warm and fuzzy. Some, like Alexander Aitken, could recite the value of π to 300 places. Others, like Hungarian Paul Erdos spent his life wandering the globe, seeking out mathematicians to work on unsolved problems.

Then there was the great logician Kurt Gödel, who in later years was convinced that he was being poisoned and, refusing to eat, essentially starved himself to death. And of course there was Theodore Kaczynski, the Unibomber, who although mathematically gifted, became mentally unstable and from 1978 to 1994, sent bombs to government agencies in his quest to free people from the tyranny of progress.

The Italian mathematicians Luca Pacioli (1447-1517) an Italian friar in the Catholic church and a seminal contributor to the field of accounting who introduced the double-entry system of book keeping. On the other had the French mathematician André Bloch (1893-1947) ended his life in a mental institution after killing his brother, aunt and uncle, to rid, as he confessed later, to eliminate branches of his family affected by mental illness.

Then there were mathematicians whose primary profession was not mathematics, such as George Boole who was a primary school teacher, Pierre de Fermat who was a lawyer, Marin Mersenne a theologian, and on and on.

Although we have received many candidates for the "greatest mathematician" honor, according to a large, distinguished, impartial jury of one, the following candidates must be considered.

Italian Gerolamo Cardano (1501-1576)

Cardano

Italian Gerolamo Cardano is considered by many historians to be both the greatest physician and mathematician of the European Renaissance. Although he had an abrasive personality and was generally disliked, he played a major role in the development of biology, medicine, philosophy, astronomy and mathematics.

Today he is remembered mostly for his seminal treatise *Ars Magna*, the greatest mathematics book of the Renaissance and a ground-breaking volume on algebra In this book, Cardano presents solutions of the cubic and quartic equations as well as the first systematic use of *two new kinds of numbers*, which were beyond the pale of human comprehension at the time, *negative* and *complex numbers*.

He also played an important role in the introduction of probability and predicted the exact date of his death, although some have said he committed suicide just so his prediction would come true. That is why Gerolamo Cardano is a candidate for ... *The Most Interesting Mathematician in the World.*

Swiss Leonard Euler (1707-1783)

Leonard Euler

Swiss Leonard Euler (OY-ler) was one of the greatest mathematicians of all time, having published over 800 papers covering every known branch of mathematics as well as mechanics, fluid mechanics, optics and astronomy. He was one of the founders of pure mathematics and wrote influential mathematics textbooks on calculus, introducing much of the notation which mathematics students know today, including the notation $f(x)$ for functions, the symbol $\sum$ for summation, the letter i for $\sqrt{-1}$, as well as Euler's constant e for the base of the natural logarithm. Wise people have said that the closest proof of the existence of God lies in Euler's equation

$$e^{i\pi} + 1 = 0$$

linking the five most important numbers of mathematics 0, 1, π, i and e in a single unifying equation. These contributions are even more impressive considering the fact that over the last three decades of his life, he was *totally blind*. That is why Leonard Euler is a candidate for ... *The Most Interesting Mathematician in the World.*

Hungarian Paul Erdős (1913-1996)

Paul Erdős

Hungarian Paul Erdős (AIR-daish) was so devoted to mathematics that he never married, lived out of a suitcase, and covered the globe, popping up on mathematicians' doorsteps, offering them the opportunity to work with him on mathematical problems. He was arguably the most prolific mathematician of the 20th century, having published some 1500 papers with more than 450 collaborators. His research falls into several categories, some of which he created, but most in the general area of discrete mathematics. He defied the conventional wisdom that high-level mathematical research is a young-person's game, making major contributions to mathematics until his death at the age of 83.

He once visited a university where a problem was written on the blackboard on an area of mathematics of which he was unfamiliar. Two local mathematicians had just come up with a 30-page solution of the problem and were quite proud. Erdős looked at the board and said, "Is that someone's problem?" He then went up to the board, asked a few questions and quickly wrote down a two-page proof. That's why Paul Erdős is a candidate for ... *The Most Interesting Mathematician in the World.*

——¤¤¤ΞΞΞ¤¤¤——

Hypatia of Alexandria (355-415)

Hypatia

Hypatia (*hy-PAY-shee*) **of Alexandria** was in her day the world's leading mathematician and astronomer, the only woman for whom such a claim can be made. She was respected for her virtue and was a leader of the *Neoplatonic Academy* in Alexandria, where she imparted the knowledge of Plato and Aristotle.

However her scholarship and depth of scientific knowledge was taken as witchcraft and paganism to early Christians, and one day a Christian mob tore Hypatia from her carriage, dragged her into a church where they stripped her naked and flayed her alive with shards of broken pottery. Christian thinking, today, in a more enlightened era, she is taken as a symbol of Christian virtue for her kindness and love of learning. That is why Hypatia of Alexandria is a candidate for ... *The Most Interesting Mathematician in the World.*

ΠΨΞθΦΧ

9

WAS 1, 2, 3 PRESENT AT THE BIG BANG?

When Pythagoras of Samos scratched out his famous equation on a scrap of papyrus 2500 years ago, did he conclude

> "Χολη Απαλλω, look what I've found,"

or

> "Χολη Απαλλω, look what I've created."

Maybe Pythagoras made no reference to Apollo, but he did raise the age-old philosophical conundrum of Platonism versus Intuitionism, or what most people are apt to call, how many angels can sit on the head of a pin?

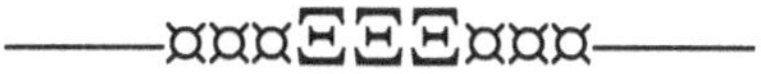

Platonism

But, if you like to dabble in such enigmas, the (mathematical) Platonist philosophy holds the belief there is a mathematical world outside our world of space and time, where all mathematical truths reside. It is a world, independent of humankind, where mathematical entities and theorems dwell. Adherents of this philosophy are called mathematical Platonists, and when a mathematical Platonist proves a theorem or makes any kind of mathematical discovery, the person believes he or she has *discovered* (not created) a truth that *already in existence* in that distant world of mathematical truths. If Pythagoras had never discovered his famous theorem, it would still be out there in the cosmos of undiscovered mathematical entities.

The great Hungarian mathematician Paul Erdos (1913-1996) often referred to "The Book," where God keeps the proofs of mathematical theorems — a Platonist in the true sense of the word.

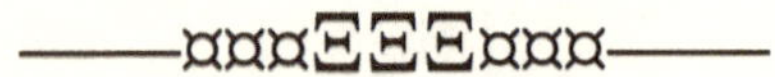

Intuitionism

On the other hand, mathematical intuitionism, introduced by the Dutch mathematician L.E.J. Brouwer (1881-1966) is based on the philosophy that mathematical ideas do not exist independent of humans, but are created as a result of human mental activity In other words, the Pythagorean Theorem did not exist before being created by Pythagoras.

The intuitionist view of mathematics has some far-reaching implications when it relates to how mathematics is practiced. One consequence has to do with the *principle of the excluded middle,* one of the three Aristotelian laws of logic. The principle (or law) states that for any proposition, either the proposition is true or the proposition is not true, nothing else. Although few would question the validity of such a claim, for an intuitionist, the claim is not universally accepted, especially when the proposition in question relates to infinite sets.

For example, the principle of the excluded middle is a basic tool used by mathematicians when proving theorems by contradiction. When proving the validity of a proposition by contradiction, one assumes the proposition is *not* true (i.e. false), then if one arrives at a contradiction of some type, one concludes one cannot assume the proposition false, and by appealing to the principle of the excluded middle, one concludes the proposition is true. But intuitionists disallow the principle of the excluded middle, and so proofs by contradiction are considered invalid.

Another characteristic of intuitionism is its non acceptance of the infinite. Infinite sets, like the natural numbers $1, 2, 3, \ldots$ or real numbers, are not considered valid mathematical entities since, unlike finite sets. Intuitionists claim that infinite sets cannot be constructed, only *imagined*. An intuitionist would say the natural numbers are "potentially infinite" since they are unending, but intuitionists do not consider the "completed set" as a legitimate mathematical entity.

Although most mathematicians consider themselves Platonists, artists and composers consider themselves intuitionists since they imagine themselves "creating" art and music from nothing, not discovering hidden scores and paintings already existing in some mythical world of art and music.

Although Platonism and intuitionism are distinct philosophies, people often have leanings for both philosophies. Imagine the philosophical dilemma of a sculptor staring at a block of marble. On one hand, the sculptor might think like a Platonist and imagine Michelangelo's *Pieta* is inside, waiting to be released by chopping off the stuff around it. On the other hand, the sculptor might view the block as an intuitionist, imagining the creation of a new work of art, which he will baptize, the *Pieta*.

The debate whether mathematical truths are discovered as a Platonist claims, or invented as an intuitionist claims, has been with us since the time of the Greek philosopher Plato, the originator of the Platonic philosophy. Today, if you do ask the average mathematician which philosophy he or she subscribes, they are apt to say they haven't given it a great deal of thought, but pressed will probably admit to being a Platonist. This belief is also shared by most of the great mathematicians of the past, including Newton, Gauss, Hardy, Cantor, and Gödel.

ΠψΔΦθΛ

10

MACHO BRAGGADOCIO DEBUNKED BY MATHEMATICS

There are things so ridiculously obvious no one gives them a second though. But when those ridiculously obvious things are restated in a different context, well, things can be ridiculously non-obvious.

Imagine four boys and four girls, getting together for a social evening of song and dance.

The next day each is asked with whom they danced the previous night. The following diagram gives a visual display of their responses.

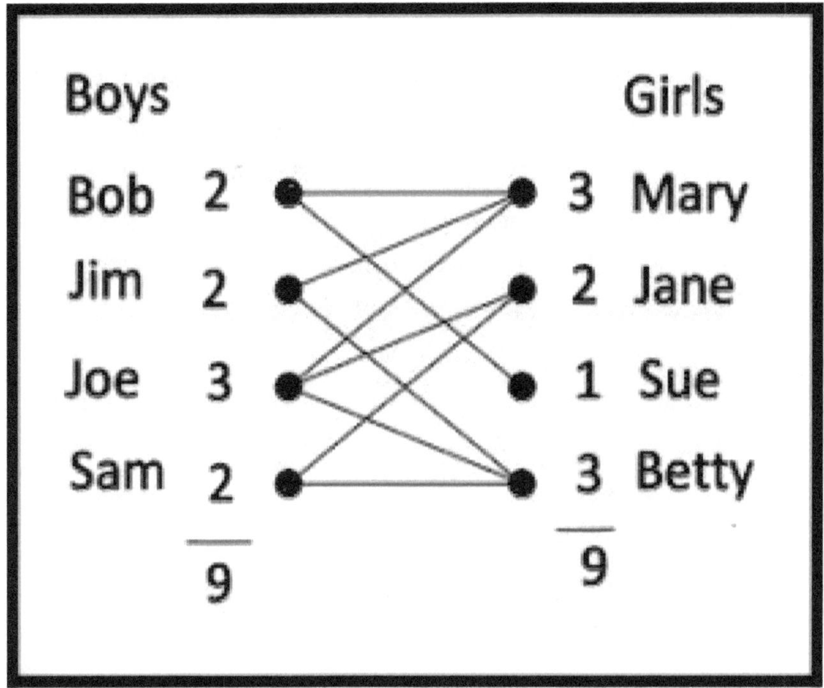

Social relations

For example, Bob danced with Mary and Sue, Jane danced with Joe and Sam, and so on. The number of dance partners for each person is displayed next to their names. The four boys danced a total of 9 times, which if you believe a dance involves one boy and one girl, is also 9 for the girls. Hence, the average number of dance partners for both boys and girls is

$$9/4 = 2.25$$

Like I said, some things are so ridiculously obvious no one gives them a second thought.

———¤¤¤◘◘◘¤¤¤———

With that in mind, let's move on to a 2002 news article in the *Bloomberg News*.

> *A US government survey conducted by the Centers for Disease Control and Prevention discovered that the average number of sexual partners was 6.8 for men and 3.7 for women. The survey included 6,237 men and women with the ages from 20 to 59, who were asked about their sexual habits.*

The diagram below shows hypothetical sexual liaisons in the polled population, where the number alongside each individual gives the number of liaisons. Note, although the number of males and females polled may not be exactly the same, and the liaisons may not be within the group polled, the outcome is (basically) the same as for the dance partners for the 4 boys and 4 girls.

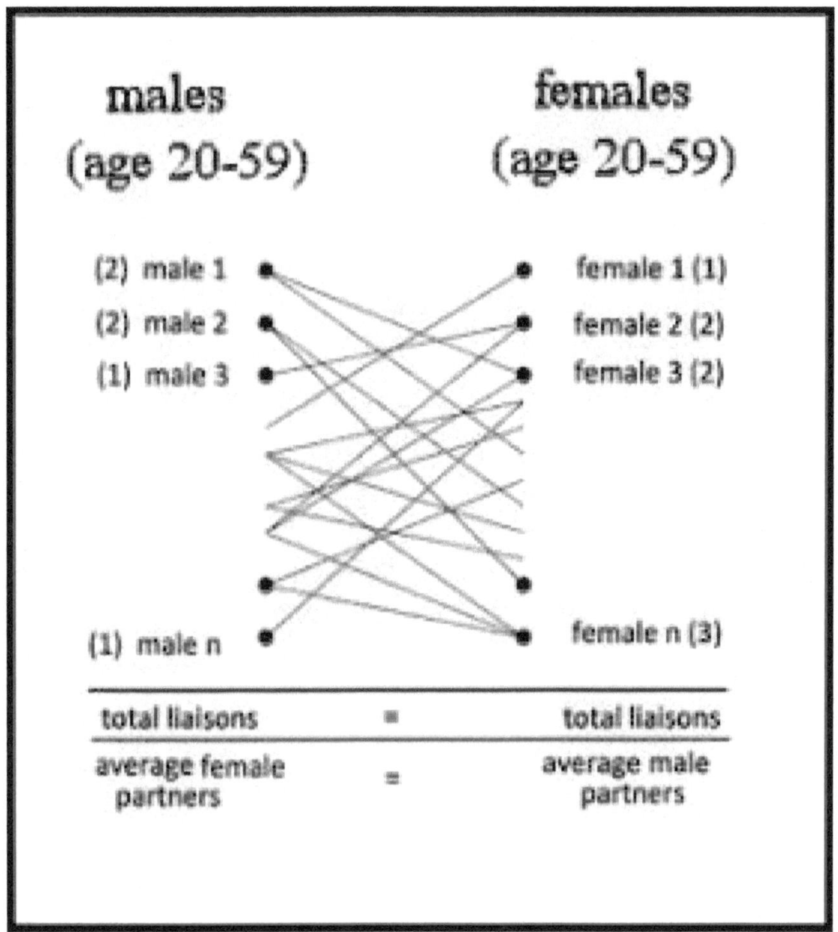

Social liaisons

In the Bloomberg survey, the number of males

and females was not *exactly* the same, and the individuals being polled had sexual partners both inside and outside the polling population, so the comparison with the 4 boys and 4 girls is not completely identical. However, the population polled constituted a homogenous group insofar as sexual mores, so the fact individuals had relations outside the population is irrelevant.

——¤¤¤ΞΞΞ¤¤¤——

So now, we get to the burning question, and arguably, the more interesting one. Why did the survey indicate males have almost twice the number of sexual partners as females? We leave it to the reader to field that one.

ΠΨΞθΦΧ

11

STREAKING FOR MATH AWARENESS MONTH

Pardon me for the publisher's proofreading lapse. The title of this story is supposed to read *Shrieking* for Math Awareness Month. You see, I get all exercised every April when Math Awareness Month rolls around since at that time newspapers are gorged with no-account activities like who won an Oscar, who's going to win a

Grammy, and the NFL draft and who the Jets are going to pick for that gaping hole at linebacker.

It is for that reason that I would like to propose the following quiz, asking you, dear reader, how many of the eight following names you recognize: Tom Brady, Johann Radon, Taylor Swift, Yves Meyer, Johnny Depp, Andrei Markov, Lebron James and Isaac Schoenberg.

Actually, my quiz is a fraud. It's just my awkward way of promoting *Math Awareness Month*, and to make the point that although mathematicians and mathematical discoveries never make headlines like winning a Super Bowl or a Granny, they might be more important in our daily lives, and that leads to the four unrecognizable names on your list; Johann Radon, Yves Meyer, Andrei Markov, and Isaac Schoenberg.

Johann Radon

Let me tell you about Johann Radon, probably the first person on your unfamiliar list. Johann

Radon was an Austrian mathematician who in 1917 worked out a mathematical theory for reconstructing the shape of an object by looking at its shadow from different directions. It's a little bit like finding out where all the nuts and fruit are inside a fruitcake by looking at slices of the cake.

Radon's ideas went unnoticed until the 1970s when an English engineer and a physician from Tufts University used Radon's theory in the invention of the CAT Scan. To them the fruitcake was the human body, the nuts and fruit were internal organs or maybe a tumor. To produce a picture that shows the exact location of things, the CAT scan takes X-rays from many different angles, then use Radon's formulas (called the inverse Radon transform) to reconstruct the picture.

Yves Meyer

Let's now move on to person #2 on the list where no doubt you said under your breath, "Who the devil is that?" Thirty years ago French mathematician, Yves Meyer, was instrumental in the introduction of the wavelet, a mathematical theory with applications in all areas of science and technology, one is the compression of television pictures, digital camera pictures, jpg pictures in your computer, mp3 digital music, and so on. If these files can be compressed without losing any resolution, then they can be transmitted faster, either by television transmitters or over the Internet.

The basic idea behind wavelets is that if a picture consists of only one color, there is no reason to send every pixel or dot; just sending the color is sufficient. It's never that simple of course since pictures are always a complex mixture of colors, but in Meyer's theory, the different colors and hues are collected by a complex process into wavelets and these wavelets are transmitted rather than transmitting every pixel. Wavelets are used by the FBI today to compress the

200,000,000,000,000,000

bits of data required to store the fingerprints of convicted felons.

Why it is so many people are turned off by mathematics? Someone said it's a result of the "piano syndrome." Learning to play the piano and learning elementary mathematics are a lot alike. In both cases the beginner must pass through a rigid orientation. The beginning pianist spends months developing finger dexterity by playing scale after scale.

Only after the basics are mastered can a student of the piano interpret a Chopin concerto or the student of mathematics use one's imagination to build mathematical worlds. Most beginning students of mathematics do not have a good idea of what mathematics is about and most do not survive the orientation to see the beauty of the subject and the rewards that follow.

Andrei Markov

Let me now tell you about mystery person #3

on the unidentified list. Andrei Markov was a Russian mathematician who in 1906 worked out a mathematical theory, called Markov Chains, for describing how many physical systems evolve over time. Such a system might be anything from a baseball game, a frog jumping from lily pad to lily pad, the evolution of a biological population like bacteria or viruses, or even a person surfing the internet, clicking from one webpage to another.

Although Markov did not have any specific application in mind for his theory, today Markov Chains are used by engineers and scientists the world over. When Google decided upon a strategy for ranking webpages, they imagined a person starting at some webpage, then moving from page to page. This can lead to dead ends at pages that have no outgoing links or around endless clicks of interconnected pages.

This type of random walk is called a Markov Chain and using Markov's theory, it's possible to find the fraction of time the surfer will spend at each page. Google uses Markov Chains to compute the *PageRank* of each webpage. A webpage has a high PageRank if it has links from other pages of high rank. If you enter the keyword "mathematics" in the Google search engine, the webpages that show up will be ones with the highest PageRank for that keyword. Markov, who died in 1922, would be amazed at how his discovery is being applied today.

———¤¤¤¤ΞΞΞ¤¤¤¤—

Isaac Jacob Schoenberg

So, who is #4? In the 1940s while a professor of mathematics at the University of Pennsylvania, Isaac Jacob Schoenberg developed the theory of what are called "spline curves."

Splines are special kinds of curves and surfaces that have a nice appearance, are flexible, and most importantly, can be manipulated in a computer. His early work with splines is one of the cornerstones in the broader area of computer graphics.

Automobile draftsmen for Ford and General Motors create new car designs by means of spline surfaces drawn by a computer. Figures which are drawn in animated movies, such as *Shrek* or the *Lion King*, also owe a lot to the invention of splines. The next time you go to an animated movie you may think you're watching *Shrek*, but you may be watching one of Schoenberg's splines generated by a computer.

ΠΨΞθΦΧ

12

A REMEDY FOR ALWAYS BEING PICKED LAST IN GYM CLASS

There is an old song *I'm Good at Being Bad,* which if it had been around in my youth, would have been my gym-class mantra. You see, I was the proverbial APLIGC, meaning the kid *always picked last in gym class.* I was the deemed-hopelessly-athletic individual in the back always chosen last for everything from Dodge Ball to Steal the Bacon.

Using my traumatic background as motivation, I would like to suggest a few apt words of advice to my fellow castaways, who also never go high in the athletic-draft. The next time teams are chosen and your number has not been called, you politely ask the chosen ones, "Just exactly how many ways *can* ten players can play 5-against-5? Is it in the millions?

What do you think?

———¤¤¤ΞΞΞ¤¤¤———

To understand the number of ways 10 players can play five-against-five, it is necessary to understand the concept of a combination, which is nothing more than a collection of things, be it a collection of five basketball players or a collection of the three letters *a,b,c.*

An illustration of combinations of various sizes $p = 1,2,3$ selected from the set .of three objects is shown in Table 1.

$r=1$	$r=2$	$r=3$
$\{a\}$	$\{a,b\}$	$\{a,b,c\}$
$\{b\}$	$\{a,c\}$	
$\{c\}$	$\{b,c\}$	

Table 1: Combinations of size 1, 2, and 3

Observe in Table 1 that there are three combinations of size $r = 1$, three combinations of size $r = 2$, and one combination of size $r = 3$.

But we are not so much interested in listing these combinations like we have in Table 1, but simply want to know how *many* combinations there are of a give size. Fortunately, there is a simply formula that determines the number of combinations of a given size and it is called the binomial coefficient and is given by the following formula.

The number of combinations of size r taken from a larger set of size n is binomial coefficient $C(n,r)$, and is computed from the formula

$$C(n,r) = \frac{n!}{r!(n-r)!}$$

where $n! = n\,(n-1)(n-2)\ldots(2)$ and is called n factorial. The numbers $r!$ and $(n-r)!$ are computed in a similar way.

Using the above formula, we can determine the number of ways 10 players can choose sides to play 5-against-5. Imagining yourself as one of the 10 players, simply yourself how ways can you can choose 4 teammates from the 9 players. The number of ways is simply the number of combinations of size 4 selected from a set of size 9, which is the binomial coefficient:

$$C(9,4) = \frac{9!}{4!5!} = \frac{9{\cdot}8{\cdot}7{\cdot}6{\cdot}5{\cdot}4{\cdot}3{\cdot}2}{(4{\cdot}3{\cdot}2)(5{\cdot}4{\cdot}3{\cdot}2)} = \frac{9{\cdot}8{\cdot}7{\cdot}6}{4{\cdot}3{\cdot}2} = 126 \text{ ways}$$

In other words, there are 126 different ways 10 players can play five-against-five.

We can also determine the number of ways 4, 6, and 8 players can play 2-against-2, 3-against-3, and 4-against-4 respectively. This is done similarly to how we did for 10 players.

Number of ways 4 players can play 2 on 2

$$C(3,1) = \frac{3!}{1!2!} = \frac{3 \cdot 2}{2} = 3$$

Number of ways 6 players can play 3 on 3

$$C(5,2) = \frac{5!}{2!3!} = 10$$

Number of ways 8 players can play 4 on 4

$$C(7,3) = \frac{7!}{3!4!} = 35$$

Number of ways 10 players can play 5 on 5

$$C(9,4) = \frac{9!}{4!5!} = 126$$

Number of World Series

If the above problems do not humble the gym-class captains, you might resort to baseball and ask for the number of ways can two baseball teams play a 4, 5, 6, or 7 game World Series. The answer again depends on combinations.

Suppose a World Series lasts six games with results WLLWWW, where W represents wins for the ultimate winner and L represents losses. Since the series winner in a 6-game series always wins game 6 and 3 of the previous 5 games, we conclude:

> *The number of 6-game World Series is the number of combinations of size 3 taken from a set of size 5.*

That is

$$C\left(5,3\right)=\frac{5!}{3!2!}=\frac{5\cdot4\cdot3\cdot2}{(3\cdot2)2}=10$$

By the same reasoning, the number of 4, 5, 6 and 7-game series are

Number of 4 game World Series

$$C\left(3,3\right)=\frac{3!}{3!0!}=\frac{3\cdot2}{3\cdot2}=1$$

Number of 5 game World Series

$$C(4,3) = \frac{4!}{3!1!} = \frac{4 \cdot 3 \cdot 2}{(3 \cdot 2)2} = 4$$

Number of 6 game World Series

$$C(5,3) = \frac{5!}{3!2!} = \frac{5 \cdot 4 \cdot 3 \cdot 2}{(3 \cdot 2)2} = 10$$

Number of 7 game World Series

$$C(6,3) = \frac{6!}{3!3!} = \frac{6 \cdot 5 \cdot 4 \cdot 3 \cdot 2}{(3 \cdot 2)(3 \cdot 2)} = 20$$

Adding up the number of 4, 5, 6 and 7 game series, we see the total possible number of World Series is 35. One 7-game World Series was WLWLLWW, played in 1937 when the St. Louis Cardinals beat the Detroit Tigers. A famous WWWW 4-game sweep occurred in 2004 when the Boston Red Sox swept the St. Louis Cardinals, ending a 86 year drought.

You can now ask your gym-class tormentors which World Series outcomes, be they be WWLLWLW, LWWWW were most frequent and least frequent in World Series history, which began in 1903? I will tell you half the answer, the most-played series is a 4-game sweep

WWWW, which has occurred 32 times between 1903 and 2016. There are five series that have *never* been played. What are they?

ΠΨΞθΦΧ

13

ANTIPODAL SOUL MATES IN CELSIUS

The person who said a picture is worth 1000 words was certainly not a mathematician, else he would have said 10,000. That ancient proverb comes to mind when seeking a verification of The Meteorologist's Problem:

At any instant of time, there are two opposite

points on the surface of the earth that have the same temperature.

To understand why this seemingly 'out-there' statement is not so 'out-there,' take a point on the surface of the earth, along with its antipodal point, the antipodal point being the point diametrically opposite the given point. Another way of thinking about antipodal points is that they are connected by a straight line passing through the center of the earth. The first antipodal points that come to mind are the North and South Poles, although Leon, Spain and Wellington, New Zealand would work as well.

The antipodal point for any point on earth can be found from the following map. Note that the antipodal point for any place in the U.S. is somewhere in the Indian Ocean. (Hence, debunking the old mother's adage of telling their kids to stop digging in the backyard, lest end up in China.)

Imagine the great circle on the surface of the earth passing through the North and South Poles and along the 90 degrees East and 90 degrees West Longitude lines. See Figure 1. If the temperatures at the North and South Poles are the same, then the claim there are always antipodal points with the same temperature is proven. If the temperatures at the poles are not the same, assume the temperature at the North Pole is +20 degrees and -20 degrees at the South Pole. See Figure 1.

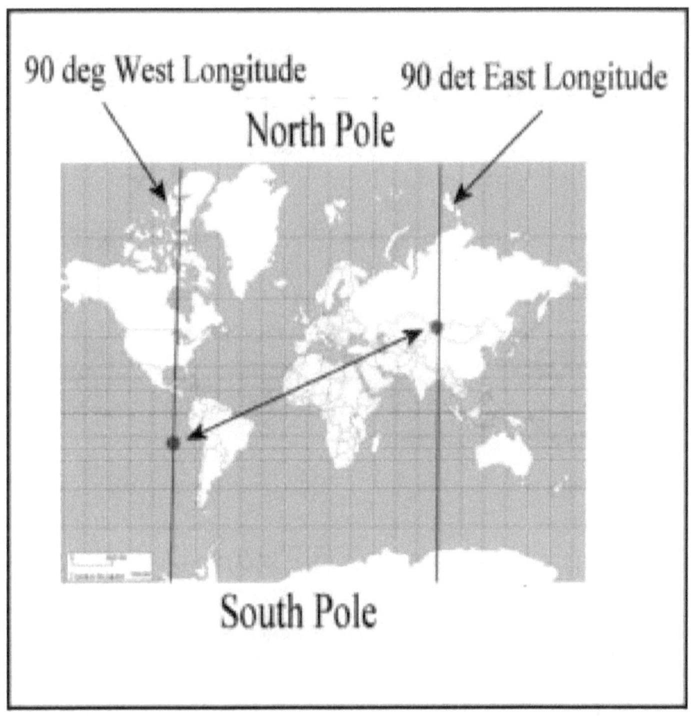

Figure 1 Antipodal points with equal temperatures

Since temperature changes continuously on the

surface of the earth as we move around the globe, the temperature difference between antipodal points on the East Longitude and West Longitude paths changes continuously from +40 to -40 degrees (or from -40 to +40) and hence there exists antipodal points where it is 0, which means the the points have the same temperature. See Figure 2.

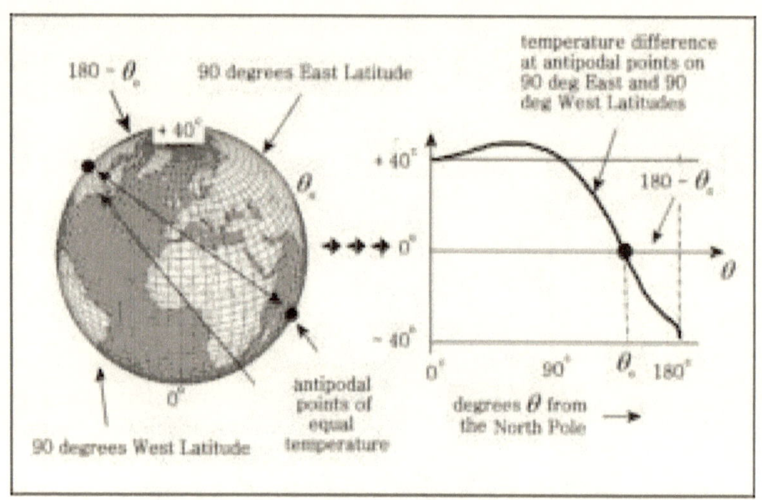

———¤¤¤ΞΞΞ¤¤¤——

The visualization this temperature property on a sphere is a result of a well-known theorem in mathematics called the Intermediate Value Theorem, which states that if a continuous function is defined on a closed interval $[a, b]$ and if the value of the function is negative at a and positive at b, then at some point between a and b the value of the function is zero.

PS: The Meteorologist's Problem is closely related to the *Hiking Monk Problem*, where a monk leaves a base camp at dawn, hikes up a twisty trail to the top of a mountain, arriving at dusk, where he camps overnight. The next morning the monk departs camp at the same time as the day before and hikes down the same twisty trail, when all of a sudden he looks at his watch and exclaims, "I was at this exact spot on the trail at exactly the same time yesterday." Now that you are familiar with the Meteorologist's Problem, the solution of the Hiking Monk Problem should be a snap. Just remember, a picture is worth a thousand words.

ΠΨΞθΦΧ

14

MATHEMATICAL ABRACADABRA IN 3-CUPS

I really shouldn't spill the beans. Oh well, why don't I spill the beans. After all, I'm not a magician. What could they do to me other than saw me in half, tie me in chains and lower me headfirst into the East River. Anyway, if mathematicians were half as secretive and close-mouthed about mathematics as magicians are about their craft, we'd still be using stone axes and drawing wild animals on cave walls.

It may come as a surprise, but there are magic tricks that actually use mathematics and logic. One of the more impressive ones was invented by Robert Hummer, a magician from Havre de Grace, Maryland, who called his trick the Three-Cup Trick. The trick not only leaves people bewildered, but to understand it involves an exercise in (horror of horrors) logical thought. It has even been said that some magicians who perform the experiment are not 100% sure themselves why it works.

The trick: Three cups are placed on a table upside down as shown in Figure 1. The positions (not the cups) are numbered 1, 2, and 3 as viewed left to right by a spectator.

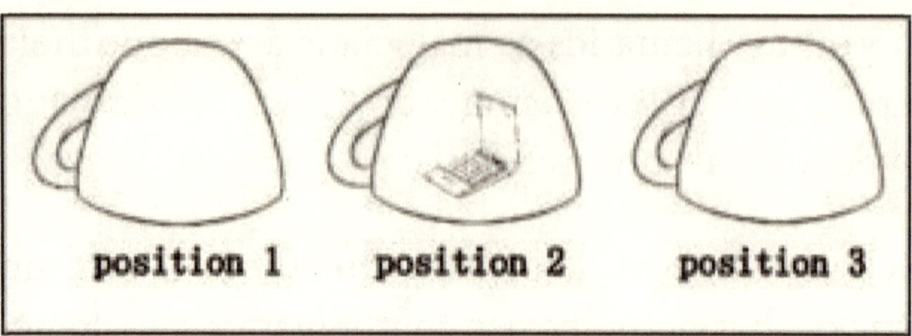

Figure 1

The Trick: The trick begins when you (the magician) turn your back to the table and ask a spectator to place a small object, say a matchbook, under one of the cups.

Step 1: Preliminary exchange You then ask the spectator to exchange the two cups *not* containing the hidden object without telling you.

Step 2: Major exchanges You now ask the spectator to begin scrambling the cups, exchanging them in pairs, each time calling out the two cups being exchanged. For example, if cups in positions 2 and 3 are being exchanged, the spectator says, "Exchange 2 and 3." Don't forget, the numbers refer to the *positions* of the cups, not the individual cups. The spectator continues exchanging cups as long as he or she wishes (maybe 3-4 exchanges). Also, when two cups are exchanged, the cups should be slid across the table, not picked up revealing the hidden object.

Step 3: Voila! After the spectator stops exchanging cups, you (the magician) turn around and lift the cup containing the hidden object.

You tell the spectators the trick works because you have a special eye that can see through cups. If they don't buy that, you make up another reason.

——¤¤¤☰☰☰¤¤¤—

Determine the Position of the Hidden Object

Although the spectators do not distinguish among the three cups, you (magician) designate one of the cups to be the 'marked cup.' If the cups look alike, you must find some minor feature on one of the cups. Before turning your back to the table, note the position, 1,2 or 3 of the marked cup. It might be easiest to place the marked cup in position 1, although you should appear to be placing them at random.

——¤¤¤ΞΞΞ¤¤¤—

Preliminary Exchange

After turning away from the table and asking a spectator to place a small object under one of the cups, you then ask the spectator to carry out a preliminary exchange by exchanging the two cups *not* containing the hidden object (of course without revealing to you which two cups are being exchanged). Although this step is critical, you can disguise its importance by simply saying it's a practice step. (Maybe make a joke saying the spectator needs some practice.) After the trick is performed, the spectator will probably forget the step.

——¤¤¤ΞΞΞ¤¤¤—

Further Exchanges

After the spectator makes the preliminary exchange, you begin with one of your thumbs pressed against one of three fingers, corresponding to the position of the marked cup. If the marked cup is initially in position 2, press your thumb against finger 2, and so on. Then, when the spectator begins exchanging cups (not the preliminary exchange) and calling out the cups being exchanged, keep track of the marked cup. If the spectator says, exchange 2 and 3, you move your thumb from finger 2 to finger 3. From then on, for each exchange, track the position of the marked cup. When the game is completed, your thumb will point to one of the three positions.

Finding the Hidden Object

When the spectator stops exchanging cups, suppose your thumb points to finger 2. You then turn around and inspect the cups and make the startling following announcement:

> *If the marked cup is at position 2, the same position as your thumb, then this cup contains the hidden object. If the marked cup is* not *in position 2, the position of your thumb, then the hidden object*

is under the unmarked cup not in position 2 (the position of your thumb).

——¤¤¤ΞΞΞ¤¤¤—

——¤¤¤ΞΞΞ¤¤¤—

Logic Behind the 3-Cup Trick

The trick can be understood by demonstrating how it works in a couple different situations.

Case 1: Hidden object under marked cup

Assume the marked cup, decorated with stars, is initially placed in position 1, while the hidden object, a matchbook, is hidden under the marked cup as shown in Figure 2.

The preliminary step, consisting of exchanging the cups not hiding the matchbook, will exchange the cups in positions 2 and 3, leaving the marked cup and its hidden matchbook unchanged. After that, the spectator starts exchanging all three cups while you track the marked cup with your thumb.

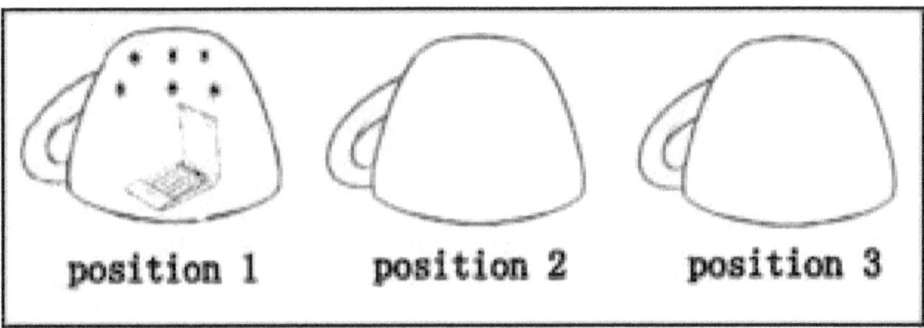

Figure 2

When this process is finished, the marked cup will be in the *same position* as predicted by your tracking. In other words, if your thumb points to the final location of the parked cup, the hidden object is under the marked cup.

Case 2: Hidden object not under unmarked cup

Assume the marked cup is again initially placed in position 1, but now the matchbook is hidden under one (either one) of the two unmarked cups shown in Figure 3.

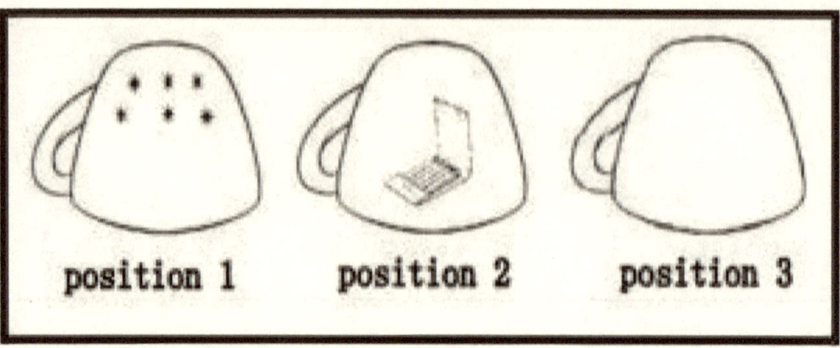

Figure 3: Marked cup in position 1

Now, on the preliminary step, the spectator exchanges cups 1 and 3 (cups not containing the hidden object), resulting in the new arrangement illustrated in Figure 4.

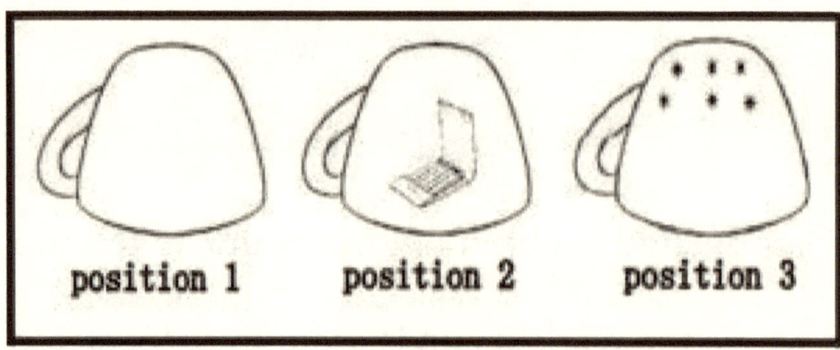

Figure 4: Cup arrangement after exchange 1 and 3

Since you do not move your thumb on the preliminary step, from this point on your thumb is tracking the unmarked cup that does not contain the hidden object.

Hence, after several cup exchanges, if you turn around and see that your thumb does not point to the marked cup, then the hidden object is under the unmarked cup not pointed to by your thumb.

Hence, after the conclusion of the cups being turned around, the magician finds the hidden object as follows:

1. If the magician's thumb points to the marked cup, the hidden object isunder the marked cup..
2. If the magician's thumb does not point to the marked cup, the hidden object is under the unmarked cup not pointed to by the magician's thumb.

ΠΨΞθΦΧ

15

HOW I LEARNED TO STOP WORRYING AND LOVE THE PROOF

I'd like to tell you a few things about mathematical proofs you might have overlooked back in your school days. Mathematical proofs come in all shapes and sizes. There are proofs that mathematicians call "beautiful proofs", where each step of the verification elicits a response like interesting, clever, or wow, I never saw *that* coming, and so on. On the other hand, there are yeoman type proofs that get the job done, but that's it, proofs which elicit responses like ok, ok I get it, and so on.

There are also mathematical proofs, like the proof of Fermat's Last Theorem, which states that for all positive integers the equation

$$a^n + b^n = c^n$$

never holds for integer exponents n = 3, 4, 5, ... In this case the conclusion is not really important, but the steps used by English mathematician Andrew Wiles have led to new insights in mathematics. It is not surprising that if it takes 358 years to solve a problem, the methodology required to solve it breaks new ground.

On the flip side, there are proofs whose conclusion is important, like the product of two negative numbers is positive, but the steps required to prove it result are not overly inspiring.

Some proofs contain thousands of steps,

whereas some proofs only a few. In 1769 the Swiss mathematician Leonard Euler made the conjecture that it is *impossible* to find five positive integers *a*, *b*, *c*, *d* and *e* that satisfy the relation

$$a^5 + b^5 + c^5 + d^5 = e^5$$

This conjecture was not known to be true or false for 197 years until 1966 when mathematicians Lander and Parken, with the aid of their good friend, the CDC 6600, proved Euler's claim to be false with a one-equation counterexample:

$$27^5 + 84^5 + 110^5 + 133^5 = 144^5$$

This is clearly one of the shortest proofs in mathematics history, although you might want to tell that to the CDC 6600.

———¤¤¤ΞΞΞ¤¤¤—

A clever one-line proof is one that verifies $22/7 > \pi$ and is embodied in the single equation

$$0 < \int_0^1 \frac{x^4(1-x)^4}{1+x^2}\,dx = 22/7 - \pi$$

The integral is clearly positive and hence so is its value $22/7 - \pi$. The value of the integral can easily be found by expanding the numerator, dividing by $1 + x^2$, and then integrating each term — an exercise every student of freshman calculus should be able to carry out.

——¤¤¤⊟⊟⊟¤¤¤—

There are proofs based on pictures that embody relevant ideas. One such visual proof verifies the conjecture that the sum of the first n odd numbers is n^2. In other words

$$1 + 3 + 5 + \cdots + (2n - 1) = n^2$$

Some examples are as follows:

$$\begin{aligned} &1 = 1^2 \\ &1 + 3 = 2^2 \\ &1 + 3 + 5 = 3^2 \\ &1 + 3 + 5 + 7 = 4^2 \end{aligned}$$

This result can be proven rigorously by the process of mathematical induction, although the following drawing illustrates the result nicely.

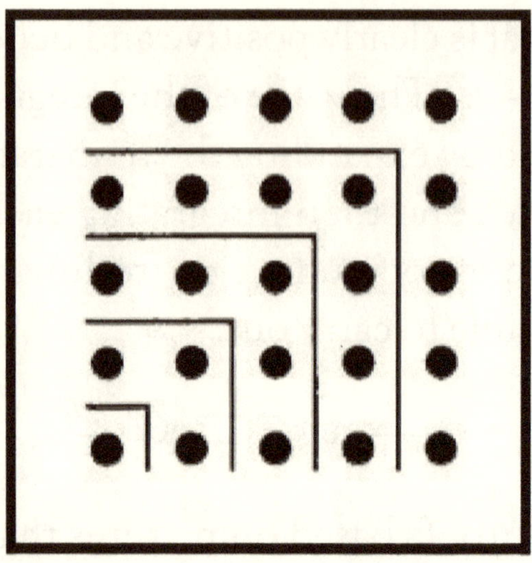

Can you "see" the identity?

However, one must be cautious when using visual proofs. The following optical illusion illustrates this point.

Do you see what's wrong here? Hint: You may think you are looking at two triangles, but if you did you would be wrong. The "hypotenuse" of both the bottom and top figures are not straight lines. The top phony hypotenuse is "bowed inward" whereas the phony hypotenuse of the bottom figure is bowed outward, allowing the one extra square to squeeze its way into the bottom figure. In order for the hypotenuse of the largesy triangle to be a straight line, the ratios of the sides of the two smaller triangles must be the same, but one ratio is 2/5 whereas the other is 3/7, clearly not the same.

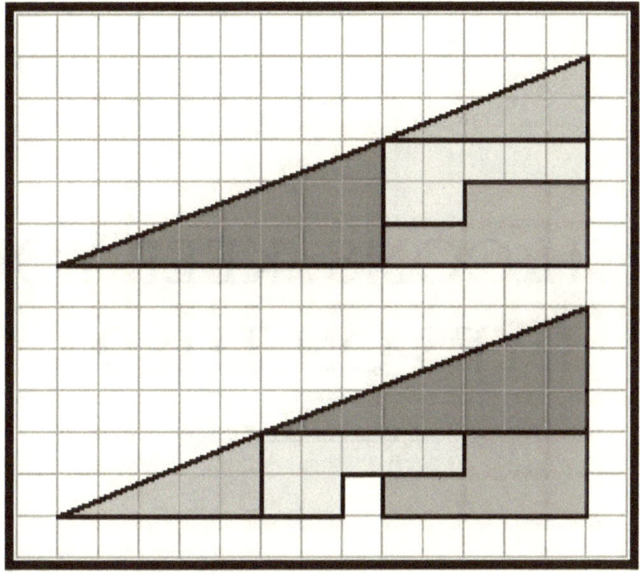

Be careful when using visual proofs

This paradox is called the Curry Paradox (or missing box paradox), named after the magician, Paul Curry.

ΠΨΞθΦΧ

16

MATH CONS: KEEP YOUR EYES ON THE X

The Great Kardini is in great form. The famous conjurer has been crisscrossing the country, wowing puzzled audiences with his otherworldly abilities. Recently, he correctly inferred a woman's age, weight, and place of birth before a mesmerized audience.

"37, 185, Boise, Idaho" he announced

So how does he do it, you ask? Actually we don't have the foggiest idea, but if it's mathematical chicanery that appeals, you've come to the right place. To assist you in becoming a mathematical hustler, we will violate the cardinal rule of the hustler and let you in on a few tricks of the trade, and although we don't teach the bending of spoons, we will show you how to perplex the mathematical unwary.

———¤¤¤ΞΞΞ¤¤¤—

Pick a Number: We begin by performing an age-old math-a-magic trick which often baffles the unsuspecting subject, after which we unwrap its inner workings.

The math-a-magician asks the subject to secretly select a number: 1, 2, 3, .. , after which the subject carries out a set of instructions arriving at a final number. At that time the magician deduces the number. Here are the rules for game. Feel free to play.

1. Pick a number 1,2,3,
2. add 10 to the number
3. add pocket change
4. multiply by 4
5. add 20
6. add 4 times your age
7. divide by 2
8. subtract twice your change
9. subtract 10
10. divide by 2
11. subtract your age
12. subtract original number
13. done, the current number is 10

No matter what initial number 1,2,3,...is chosen by the subject, the final number will always be 10. To understand this riddle, the idea is to use algebra, not arithmetic when playing the game. Suppose the secret number chosen by the subject is 21 and the subject has 25 cents in his pocket, and the age of the subject is 40. Don't use these specific numbers but choose variables, like

- n = secret number (pick $n = 3$)
- c = change (pick $c = 50$ cents)
- a = age (pick $a = 25$ years)

The following arithmetic and algebraic computations using both numbers 3, 50, and 25 and respective variables n, c, and a show that the variables are completely cancelled and the final value will always be 10.

Instruction	Arithmetic	Algebra
1 Pick a number	3	n
2 Add 10 to the number	13	$n+10$
3 Add pocket change	63	$n+c+10$
4 multiply by 4	252	$4n+4c+40$
5 add 20	272	$4n+4c+60$
6 add 4 times your age	372	$4n+4c+4a+60$
7 divide by 2	186	$2n+2c+2a+30$

8 subtract twice your change	86	$2n+2a+30$
9 subtract 10	76	$2n+2a+20$
10 divide by 2	38	$n+a+10$
11 subtract your age	13	$n+10$
12 subtract original number	10	10
13 Done, the number is 10	10	10

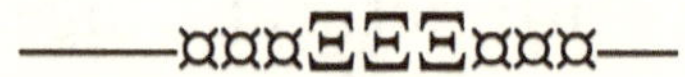

Advanced Magic

We now move on to advanced sorcery, which involves properties of decimal representation of numbers. The game consists of a person thinking of his or her age and the number of siblings. Then after performing the steps as instructed by Kardini, the person tells Kardini the final number (252 in this case), whereupon Kardini tells the person their age and the number of siblings. Here are the instructions, sample ages and siblings, and the algebraic equivalent of the numbers.

Direction	Example	Algebra
1. Enter age	25	$10a + b$
2. Multiply age by 2	50	$20a + 2b$
3. Add 10	60	$20a + 2b + 10$
4. Multiply by 5	300	$100a + 10b + 50$
5. Add # of siblings	302	$100a + 10b + 50+c$
6 Subtract 50	252	$100a + 10b+c$

The final algebraic expression

$$100a + 10b + c$$

is simply a fancy way of writing the three-digit number *abc*, where *ab* is the person's age and *c* the number of siblings. In this example Kardini says the age of the person is 25 and the person has 2 siblings. Kardini must assume the person is under 100 years old and the number of siblings less than 10.

ΠΨΞθΦΧ

17

YOU CAN'T MAKE THIS STUFF UP, FOLKS

Oh boy!

In 1999, no not 1299 silly, the state of Mississippi, or to give the benefit of the doubt to the good citizens of Mississppi, the *state legislature* of Mississippi decided to remove fractions and decimal point from the mathematics curriculum of it public secondary schools. A local newspaper reported that Mississippi possibly got its cue from the state of Kansas, when it reported:

> *... Kansas' recent measure removing the requirement for the teaching of evolution in public schools, yesterday the Mississippi legislature passed a bill eliminating fractions and decimal points from the mathematics curriculum of all public secondary schools in the state.*

"Despite the coincidental timing of the measure, this was no whim," asserted one state senator. "We had the issue under consideration for several months now," he continued.

The bill "cleared" the Mississippi Senate and thus the word when out that Mississippi secondary schools should only study whole number arithmetic and avoid those nasty fractions and decimal points.

One senator indicated that religion did play a role in the legislation stating "If cardinality is good enough for the Catholic church, it ought to be good enough for the children of the great state of Mississippi. Improper fractions' have no place in any respectable school system."

A problem with the new law arose when school librarians wondered if they could continue classifying books according to the Dewey Decimal System inasmuch as the decimal point is the foundation of the system.

Several senators indicated that an additional measure aimed at removing "irregular verbs" from English classes might be in the offing.

ΠΨΞθΦΧ

18

WHAT GREAT MATHEMATICIANS SAY ABOUT MATHEMATICS

There are so many myths about mathematics you'd think it was a religious cult — and the biggest myth of all is that there are two types of people, "math people" and "non-math people."

"I'm so bad in math, I can't even make change," is a comment you often hear someone say as a point of pride. When people ask me what I do for a living, I know it would be a mistake to say mathematics, so I just say "I'm trying to find *x*." If they tell me they teach English, I tell them "I'm so bad at reading, if it weren't for those little drawings, I couldn't tell the men's room from women's room."

Sometimes people will ask me if there are good applications of mathematics. I tell them the story of when I was working on my Ph.D thesis and trying to find applications of the Schauder fixed point theorem. I just met my future wife who knew nothing about mathematics, but I managed to explain the famous theorem to her in a language she could understand. She always nodded in agreement with everything I said so I took it to mean she understood. Later, after we were married she told me she hadn't the faintest clue what the hell I was talking about, but was so impressed with my enthusiasm that she continued to date me. So you see, sometimes you can find applications of mathematics in the least likely places.

So, why is it so many people are turned off to

mathematics? My own theory is it's the "piano syndrome." Learning to play the piano and learning elementary mathematics have a lot in common. In both endeavors, the beginner must pass through a rigid orientation. The beginning pianist spends months developing finger dexterity by playing scale after scale. In mathematics the child must first learn to count, then arithmetic, algebra, and so on. In both cases, such activities are considered by most people to be dull and monotonous. Only after the basics are mastered can the student of the piano interpret a Chopin concerto and the student of mathematics use one's imagination to build mathematical worlds. It is unfortunate that beginning students of mathematics never survive their rigid introduction to experience the exciting mathematical world that lies beyond.

What better way to throw off your old misconceptions of mathematics and learn its true meaning straight from the horse's mouth — by perusing the thoughts and opinions of the greatest mathematicians who ever practiced the Queen of the Sciences.

The art of doing mathematics is finding that special case that contains all the germs of generality. — David Hilbert

I am accustomed, as a mathematician to living in a sort of vacuum surrounded by people who declare with an odd sort of pride that they are mathematically illiterate. — David Mumford

I think it is a peculiarity of myself that I like to play about with equations, just looking for beautiful mathematical relations which maybe don't have any physical meaning at all. Sometimes they do. —Paul Dirac

The mathematician may be compared to a designer of garments who is utterly oblivious of the creatures whom his garments may fit. — Jean d'Alembert If a mathematician wishes to disparage the work of a colleague, the most effective method is to ask where the results can be applied. — Alfred Tarski

Mathematics has the inhuman quality of starlight, brilliant and sharp, but cold. — Hermann Weyl

It may be appropriate to quote a statement of Poincare, who said there must be something mysterious about the normal law since mathematicians think it is a law of nature whereas physicists are convinced that it is a mathematical theorem. — Mark Kac

It would be better for the true physics if there were no mathematicians on earth. — Daniel Bernoulli

There is no branch of mathematics, however abstract, which may not some day be applied to phenomena of the real world. — N.I. Lobachevsky

God made the integers; all the rest is the work of Man. — Leopold Kronecker Mathematical genius and artistic genius touch one another. —Gosta Mittag–Leffler

All the truths of mathematics are linked to each other, and all means of discovering them are equally admissible. — Adrien-Marie Legendre

The apex of mathematical achievement occurs when two or more fields which were thought to be entirely unrelated turn out to be closely intertwined. — Gian-Carlo Rota

The elegance of a mathematical theorem is directly proportional to the number of independent ideas one can see in the theorem and inversely proportional to the effort it takes to see them. — George Polya

If people do not believe that mathematics is simple, it is only because they do not realize who complicated life it — John von Neumann

God exists since mathematics is consistent, and the devil exists since we cannot prove it. — Andre Weil

It becomes the urgent duty of mathematicians, therefore, to meditate about the essence of mathematics, its motivations and goals and the ideas that must bind divergent interests together. — Richard Courant

In my opinion a mathematician need not preoccupy oneself with philosophy — an opinion, moreover, which has been expressed by many philosophers. — Henri Lebesgue

The mathematician does not study pure mathematics because it is useful. He studies it because he delights in it and he delights in it because it is beautiful. — Henry Poincaré

ΠψΔΦθΛ

19

THE PARENTS ARE COMING! THE PARENTS ARE COMING!

A grade-school teacher once told me that although she loved teaching children, that over the years she had to endure insults, smears, badgering, and even threats. Not by her students silly, but by (albeit, only a few) parents of the students. If you ask the average grade-school teacher they will tell you the students are the best part of teaching and their parents are the worst part. As for myself in teaching college-level students for 50 years, never once was I ever treated with anything less than respect and friendship from parents of my students.

I once knew a teacher, however, who had less than cordial relations with a few parent — the blame being shared by both parties. The following document was found in the desk of said professor.

—¤¤¤ΞΞ¤¤¤—

The day began as all good Parents' Days .... *bad.* Here I sit, perched behind my desk, waiting for them to come. They come not by ones or twos, but in swarms. Here they come. They're at my door. Come in! Come in! They come from upstate, from down-state, from all-around-the-state. They come from little towns, from big towns, from middle-sized towns. They are butchers, bakers, and candlestick makers.

This year I am ready for them, ready for another Spanish Inquisition. Should Johnny go on to graduate school for the Ph.D? What about Sally's prospects for the Nobel Prize in Human Development?

I tell them their son couldn't be admitted to an open-admission community college if they bribed the dean. The only prize their daughter could win was one for endurance, entertaining the campus rugby team — and every time their kid solves an equation, canaries sing at the South Pole. Ha! Ha! Ha! Ha! Ha! Ha!

Don't these people know who they're talking to? Don't they know I'm a research mathematician, and not their kid's babysitter? Don't they know I'm the mathematician that solved the *Schmitzheimer Conjecture*?

One woman conceded that her son Billy was probably ... only average.

"Average? Average?" I snorted. "He'd be average if he lived in a zoo." Ha! Ha! Ha! Ha! Ha! "

"The day your kid is average Sasquatch will streak the Commons. Ha! Ha! Ha! Ha! Ha!"

One woman let me know that her Isabelle was only average and would be moving on to graduate school for further studies in Sociology.

“She’s been looking at Penn State,” she mused. “She feels she’d be right at home in University Park”

“University Park? She’d be more at home in *Jurassic* Park! Ha! Ha! Ha! Ha! Ha!”

One thing I learned from Parents’ Day is that no parent has a below-average kid. Don’t even entertain the thought.

“I guess he’s, oh, about average,” a woman recently described her boneheaded son.

Another man admitted, that although his son was poor in mathematics, he said he could do the work of three men.

“Yeah, Larry, Curly, and Mo. Ha! Ha! Ha! Ha! ”

Making some phony excuse about lunching with the Dean, I drop on all fours and begin crawling towards the door.

"Slam the door," one woman shouted as they grabbed me, kicking and screaming, and propped me up behind my desk.

"I wanted to ask you about the D you gave my Grover in remedial math,",a woman waver her finger in my face. Grover had somehow convinced his mother he was capable of human thought and should have received a higher grade for what really should have been a big fat F. I tried slithering for the door again.

"Mr. Smathers and I would like you to change Kimberly's grade to an A," a well-bred woman dressed me down matter-of-factly. "She's been under so much pressure pledging the Tri Delts you know. And of course, being on the new Scarsdale diet, things just got a trifle out of hand."

"Scarsdale diet! It's not the Scarsdale diet she's on, it's the Clydesdale diet. Ha! Ha! Ha! Ha! Ha!"

"Johnny may not be your typical Einstein," a

mother recently said. “But he can do the work of three men.”

I didn’t have the nerve to reply.

ΣΨΠΞΥΦ

20

THE A, B, C'S OF A, B, C, D AND F GRADING

Methods for measuring student performance in American colleges and universities go all the way back to the 17th century, to the very earliest American schools of higher education, like Harvard (founded in 1636) and Yale (founded in 1701), who had procedures in place for measuring student performance. Often a professor would ask the class a question whereupon the class would respond in unison, or maybe an individual student would come to the front of the room and display his learning before the class.

However, much of the earliest differentiation among students was centered more on social status than on classroom performance. At Harvard, for example, students were arranged not alphabetically but by the social standing of their families. According to a 1785 diary of Ezra Stiles, the president of Yale University at the time, the selection of the Yale Valedictorian was chosen by a vote of the Senior Class.

1785: In 1785 Yale introduced a ranking system in which students were assigned to one of four categories, where *optimi* (Latin for best) was the highest ranking, followed by a second category of second *optimi* (second best), then *inferiore* (lower), and finally the lowest *pejores* (worse). At that time Yale made reference to marking grades on a scale of 1 to 4, which is probably the origin of the 4.0 grading scale we use today.

——¤¤¤ΞΞΞ¤¤¤—

1817: In 1817 William and Mary adopted a grading system based on numbers, where students were assigned one of four numbers 1,2,3, or 4, described below.

1 Student is one of the top in the class.
2 Student is orderly and generally correct.
3 Student made little improvement.
4 Student learned absolutely nothing.

A student in those days didn't go home at the end of the semester and brag about getting a 4.0 grade point average.

——¤¤¤ΞΞΞ¤¤¤—

1877: In 1877 Harvard made its contribution to grading when it began evaluating students based on the percentage of correct answers with 100% being the best. A Harvard document, dated back to 1877, shows students being placed in one of the six divisions depending on their percentage scores.

- Division 1 90% to 100%

- Division 2 75% to 89%
- Division 3 60% to 74%
- Division 4 50% to 59%
- Division 5 40% to 49%
- Division 6 below 40%

When I was a student in K-12 my school's grading issued both E and F grades, and so the six Harvard divisions could be regarded as numerical counterparts of the six letter grades A,B,C,D,E, and F. I always interpreted the grade of E as a warning shot across the bow, telling you that if you don't shape up the next shell will have your name on it.

———¤¤¤ΞΞΞ¤¤¤—

1887: The year of 1887 was a watershed year in the annals of student grading. It was that year when Mount Holyoke College made its seminal contribution to student grading with the introduction of the letters A,B,C,D, and E, five letters implanted in the psyche of every current and former student in America.

Mount Holyoke College 1887 Grading System

- A Excellent, 95% – 100%
- B Good, 85% – 94%
- C Fair, 78% – 84%
- D Barely passing, 75% = 77%
- E Failed, below 75%

A year later, Mount Holyoke modified the letters by replacing the E with an F. No doubt, after students told their parents a grade of E stood for Excellent, whereas there was no confusing the fact that a big fat F stood for a big fat Fail.

———¤¤¤ΞΞΞ¤¤¤—

I once asked my calculus students their opinion of the A, B, C, D, and F system. One student told me that at the age of 14 she was inspired by the totally tubular Valley Girl Moon Zappa, after which she became a totally *obnoxious* Valley Girl for about two years. My student offered her Valspeak expertise and gave me the following grody to the max A,B,C,D, and F grading system. I have absolutely no idea what they mean other than a sign that the end-times are near.

- A Awesome, like, ohmigod and A+.
- B Bitchin' B+ fer shur.
- C Cracked it, like totally rad dude.
- D Dudespeaked the test, Baldwin.
- E Eat my gnarly answers, Einstein.
- F Freakin' out, I like so passed.

———¤¤¤ΞΞΞ¤¤¤—

Now that the A,B,C,D, and F grading system is solidly embedded in today's academia, all the way from K-12 to graduate school, the question we should ask is, how accurately do grades measure student performance? Other than the obvious fact that the process of grading is mostly subjective, there is the ever-present elephant-in-the-room that the higher grades of A and B have lost their *mojo* in recent times. *Your A is not your grandfather's A.*

Gertrude Stein's astute quotation "*A rose is a rose is a rose*," may offer an enlightened metaphorical revelation in some worlds, but in the world of student grading, the relevant words are

"*An A is not an A is not an A.*"

In a 2010 study by Stuart Rojstaczer and Christopher Healy, published in the *Teachers College Record*, they collected historical data on letter grades from more than 200 four-year colleges and universities. Their analysis confirms what any educator worth his blue book knows; the percentage of A and B grades awarded in American colleges and universities over the past 50 years has skyrocketed as much as the average college textbook. The following table shows the percentage of A's, B's, C's, D's, and F's given in American colleges and universities over the 50-year period from 1960 to 2010.

	A	B	C	D	F
1960	15%	33%	35%	12%	5%
1970	25%	37%	25%	9%	4%
1980	28%	38%	24%	6%	4%
1990	35%	35%	20%	6%	4%
2000	40%	32%	20%	6%	4%
2010	43%	32%	16%	5%	4%

Percentage Grades Given: 1960 – 2010

Note that 43% of grades awarded in 2010 were A's, whereas the percentage of B's remained roughly the same at the expense of the shrinking percentage of C's, D's and F's. In fact, only about 10 percent of grades given in 2010 were D's and F's, far from the "bell-shaped" curve educators advertise.

Although both public and private schools were both guilty of grade inflation, private schools were more guilty. Schools that stress science and engineering tend to be stingier with higher grades than liberal arts schools.

The question then arises, do today's students

work harder and are smarter? The answers to these questions, not surprising, are no and no. In fact, one recent study showed that today's students spend significantly *less* time studying than they did in past times. Whether they are smarter, it remains to be seen.

Researchers argue that the reason behind the huge grade inflation in the 1960s was due to the fact that professors were reluctant to give D's and F's to male students since poor grades could likely result in an airplane ticket to Vietnam. In more recent times, the increase in good grades is probably due to a variety of factors; higher grades for the student which means better student evaluations for the professor; sympathy for students seeking graduate school acceptance and good jobs, and on and on. The reader can propose a few more.

Athough grading is an important part of school life, a million examples can be sited that prove good grades don't always translate into success, and that bad grades don't always predict failure. The following quote by Ronald Reagan, the 40th president of the United States, more or less, says it all:

> *"There are advantages to being elected president. The day after I was elected president, I had all my high school grades classified Top Secret."*

ΠΨΞθΦΧ

21

A FEW OF MY FAVORITE THINGS

I'd be the last person on earth to denigrate raindrops and roses and whiskers on kittens, but if you will allow I'd like to tell you about a few of my own favorite things ... just a few old mathematical hand-me-downs Julie Andrews failed to mention in the iconic song from Rogers and Hammerstein's musical *The Sound of Music.*

The Will Rogers' Principle

It's not as pastoral as raindrops and roses, but the following wry observation by Will Roger's caught my attention the first time I ever heard it. The adage is attributed to the wise social commentator Will Rogers, who famously (or infamously depending on which state you are from) said:

> *When the Okies left Oklahoma and* moved to California, they raised the average IQ of both states.

The not so hidden dig that Mr. Rogers was trying to impart is that the average Okie lies a little higher on the Stanford-Binet scale than the average Californian. To see that this principle can actually hold in some situations, consider two groups of four people each, the DUMBIES and the SMARTIES, where the IQs of persons in each groups are

- DUMMY IQs = {1,2,3,4}, AVEAGE IQ = 2.5
- SMARY IQs = [5,6,7,8], AVERAGE IQ = 6.5

—¤¤¤¤Ξ Ξ¤¤¤¤—

n-Narcissistic Numbers

The British mathematician G. H. Hardy once visited his good friend and fellow mathematician Srinivasa Ramanujan in a hospital. In Hardy's words:

> *"I remember once going to see him when he was ill at Putney. I had ridden in a taxi numbered 1729 and remarked that the number seemed rather dull and hoped it was not an unfavorable omen. 'No'; Ramanujan replied, 'It is a very interesting number, it is the smallest number expressible as the sum of two cubes in two different ways.'* "

In case the reader has forgotten, they are .

$$1729 = 1^3 + 12^3 = 9^3 + 10^3$$

The Indian mathematician S. Ramanujan was a child prodigy and was legendary for his fathomless knowledge of numbers and their properties. An early flirtation with numbers and their myriad of riches provides great entertainment as well as mental expansion for any child. The great mathematician Carl Gauss spent much of his childhood pouring over the behavior of numbers. There are many properties of numbers, not overly important to mathematics in general, but are great learning tools for receptive minds. One such class of numbers are the n-narcissistic numbers. An n-narcissistic number is an n digit number that is equal to the sum of the nth powers of its digits. For example

$$
\begin{aligned}
&3 = 3^1 && (1 - \text{narcissistic number}) \\
&153 = 1^3 + 5^3 + 3^3 && (3 - \text{narcissistic number}) \\
&1634 = 1^4 + 6^4 + 3^4 + 4^4 && (4 - \text{narcissistic number})
\end{aligned}
$$

are all narcissistic numbers, whereas

$$
\begin{aligned}
&35 \neq 3^2 + 5^2 \\
&320 \neq 3^3 + 2^3 + 0^3 \\
&4112 \neq 4^4 + 1^4 + 1^4 + 2^4
\end{aligned}
$$

are not narcissistic.

If you are open to suggestion, convince yourself that all the 1-digit numbers 1,2,..., 9 are 1-narcissistic, there are no 2-narcissistic numbers, there are exactly four 3-narcissistic numbers, as well as exactly four 4-narcissistic numbers. Narcssistic numbers for a few numbers are

n	n-narcissistic number
1	1,2,3,4,5,6,7,8,9
2	none
3	153, 370, 371, 407
4	1643, 8208, 9474
5	54748, 92727, 93084
6	548834
37	1219167218625434121569735803609966019

We leave the check of 37-narcissistic to the reader.

—¤¤¤¤⊟⊟¤¤¤¤—

Understanding Through Pictures

The old adage that a picture is worth a thousand words doesn't hold in mathematics. In mathematics a picture is worth *ten* thousand words. At first glance the following identities relating n^2, the square of a natural number n, to the sum of the first n odd numbers may appear unnatural, but when viewed from a picture, all mysteries are revealed.:

$$
\begin{aligned}
&1^2 = 1\\
&2^2 = 4 = 1 + 3\\
&3^2 = 9 = 1 + 3 + 5\\
&4^2 = 16 = 1 + 3 + 5 + 7\\
&5^2 = 25 = 1 + 3 + 5 + 7 + 9\\
&6^2 = 36 = 1 + 3 + 5 + 7 + 9 + 11\\
&7^2 = 49 = 1 + 3 + 5 + 7 + 9 + 11 + 13\\
&8^2 = 64 = 1 + 3 + 5 + 7 + 9 + 11 + 13 + 15\\
&9^2 = 81 = 1 + 3 + 5 + 7 + 9 + 11 + 13 + 15 + 17
\end{aligned}
$$

For example, the third identity

$$3^2 = 9 = 1 + 3 + 5$$

in the above list can be interpreted as computing the number of squares in the following 3×3 grid in two different ways: the first being the usual way of computing $3^2 = 3 \times 3$ and the second by adding the squares labeled a, b, c getting

c	c	c
c	b	b
c	b	a

Arithmetic identity?

Maybe you can find your own identity for 3^2 by adding up the 9 squares in the grid in a different way.

—¤¤¤⊟⊟¤¤¤—

The Carl Gauss Identity

Many statements involving numbers, like the following unnerving one disclosing the biblical Number of the Beast 666 as the sum and difference of three sixth powers

$$666 = 1^6 - 2^6 + 3^6$$

have no mathematical significance, although a numerologist might find a cloven-hoof embedded in it.

There are however arithmetic statements that have great mathematical importance, one being the following identity relating the sum of the first n natural numbers to a simple algebraic expression: :

$$1 + 2 + 3 + \cdots + n = \tfrac{1}{2}n\,(n+1)$$

Although this identity has never been given a special name, we would like to call it Gauss' Identity, since the German mathematician Carl Gauss realized this identity at an advanced age of eight.

The idea behind this identity was realized by Carl Gauss (1777-1855) when he was in primary school. The story goes that one day the teacher asked the class to add the numbers from 1 to 100, assuming the question would occupy the class long enough for the teacher to get some much-needed rest. Unfortunately for the teacher, the young Gauss wrote down the (correct) answer 5050 after only few seconds.. The teacher couldn't understand how his pupil could compute this sum in his head so quickly, but the eight-year old Gauss pointed out the problem was actually very simple. Mentally, Gauss imagined the sum could be expressed both forward and backward getting 50 terms of 101, and hence the sum is 5050.

We now replace 50 by an arbitrary positive integer n, and sum two versions of $1+2+\ldots+n$ as illustrated by

$$\begin{array}{ccccc} 1+ & 2+ & 3+\cdots+ & n \\ n+ & (n-1)+ & (n-2)+\ldots & +1 \\ \hline (n+1)+ & (n+1)+ & (n+1)+\cdots & (n+1) \end{array}$$

$$\underbrace{\qquad\qquad\qquad\qquad}_{n \text{ terms}}$$

Algebraic identity in the making

getting *twice* the desired sum. Hence

$$1+2+3+\cdots+n=\tfrac{1}{2}n\,(n+1)$$

ΨΦΞΘΣΛ

www.ingramcontent.com/pod-product-compliance
Lightning Source LLC
Chambersburg PA
CBHW021428170526
45164CB00001B/141

* 9 7 8 1 5 4 0 7 6 4 2 5 6 *